LogicWorks 4

INTERACTIVE CIRCUIT DESIGN SOFTWARE

Capilano Computing Systems Ltd.

New Westminster, Canada

ADDISON-WESLEY

An imprint of Addison Wesley Longman, Inc.

Menlo Park, California • Reading, Massachusetts • Harlow, England
Berkeley, California • Don Mills, Ontario • Sydney • Bonn • Amsterdam • Tokyo • Mexico City

Acquisitions Editor: Paul Becker
Assistant Editor: Anna Eberhard Friedlander
Production Manager: Pattie Myers
Senior Production Editor: Teri Hyde
Art and Design Supervisor: Kevin Berry
Senior Print Buyer: Judy Sullivan
Composition: Capilano Computing Systems, Ltd.
Cover Design: Yvo Riezebos Design
Text Design: Lisa Jahred
Text Printer and Binder: Maple-Vail Book Manufacturing Group
Cover Printer: Coral Graphic Services, Inc.

LogicWorks, DesignWorks, LogiMac, and MEDA are trademarks of Capilano Computing Systems Ltd.

LogicWorks was written in the C and C++ languages using the Borland C/C++ compiler (a product of Borland International of Scotts Valley, California).

Abel and MacABEL are trademarks of Data I/O Corporation of Redmond, Washington.

Macintosh is a registered trademark of Apple Computer, Inc., of Cupertino, California.

Windows is a registered trademark of the Microsoft Corporation, Redmond, Washington.

Intel and Intel MDS are trademarks of Intel Corp. of Santa Clara, California.

Many of the designations used by manufacturers and sellers to distinguish their products are claimed as trademarks. Where those designations appear in this book, and Addison-Wesley was aware of a trademark claim, the designations have been printed in initial caps or in all caps.

Camera-ready copy for this book was prepared using FrameMaker (a registered trademark of Adobe Systems Inc. San Jose, California.

Library of Congress Cataloging-in-Publication Data
LogicWorks 4: interactive circuit design software for Windows and Macintosh / Capilano Computing Systems, Ltd.
            p.   cm.
        Includes index.
        ISBN 0-201-44488-7
        1. Digital integrated circuits--Design and costruction--Data processing. 2. LogicWorks.
        I. Capilano Computing Systems, Ltd.
TK7874.65.L66   1998
621.3815--dc21   98-36517
                                                                                  CIP

Instructional Material Disclaimer
The programs presented in this book have been included for their instructional value. They have been tested with care but are not guaranteed for any particular purpose. Neither the publisher or the authors offer any warranties or representations, nor do they accept any liabilities with respect to the programs.

ISBN 0-201-44488-7 (book)   0-201-32682-5 (package)

        4 5 6 7 8 9 10—MA—02 01 00

Addison Wesley Longman, Inc.
2725 Sand Hill Road
Menlo Park, California 94025

Contents

3

User Interface 23

4
Tutorial 45

5
Schematic Editing 93

6

Advanced Schematic Editing 125

7

Simulation   149

8
The Timing and Simulator Tools 173

9
Primitive Devices 189

10

RAMs and Programmable Devices   215

11
Device Symbol Editing 229

12

Menu Reference 261

Appendix A— Primitive Device Pin Summary   387

Appendix B—
Device Pin Types 395

Appendix C—
Initialization File Format (for Windows) 399

Appendix D—
Timing Text Data Format

Preface

Welcome to LogicWorks™ and the world of interactive circuit design. As electronic systems have become more complex, operating speeds higher and custom chip technology more widespread, software tools for engineers have become an essential part of the design process. It is no longer possible for an individual engineer or a corporation to remain competitive while using pencil and paper for design. Powerful computer aided design (CAD) programs have been commercially developed to meet the increasing demands facing industry. At Capilano Computing Systems Ltd., our flagship product, DesignWorks, is used in government, industrial, and academic labs worldwide, providing users with speed, ease of use, and affordability. Many instructors want to give their students hands-on experience with CAD tools used in industry, but the high cost and complexity of most commercial CAD programs limit their use at academic institutions. In the light of this, we developed LogicWorks, the student version of DesignWorks, to be used by students in lab settings and by instructors as an interactive teaching aid. LogicWorks was created with the following goals in mind:

- To give the student an introduction to the concepts and practicalities of using CAD tools.

- To provide a "virtual workbench" that allows the student to quickly test out circuit design ideas and document results.

- To be intuitive and easy to use, so that time is not wasted on the details of installing and operating the software.

- To provide the features and interfaces necessary to work with current design technologies.

- To provide an upward path to professional design tools used in industry.

LogicWorks4 has a new and completely redesigned user interface and adds several major features not found in earlier versions. In particular automatic sheet borders and new graphics import capabilities let you create finished, professional-looking diagrams.

Acknowledgments

Many people at Capilano Computing and at Addison Wesley Longman Publishing provided invaluable help in bringing this version of LogicWorks into the world. Particular thanks go out to our editor Paul Becker and assistant editor Anna Eberhard Friedlander at AWL, and to Neil MacKenzie at Capilano, all of whom remain cheerful and fun to work with despite setbacks and schedule pressures. In addition, we would like to acknowledge our many friends and supporters in the academic and industrial worlds who continue to provide valuable feedback and support for the ongoing development of this product.

Chris Dewhurst
Vancouver, B.C., Canada
October, 1998

1
Introduction

Welcome to the world of electronics design using LogicWorks! The purpose of this tutorial/manual is to get you acquainted as quickly as possible with all the powerful editing and simulation features of the program.

LogicWorks is available for the Windows and Macintosh platforms. To make it easier for you to learn the basic functioning of the program, the first three chapters maintain separate sections for the Windows and the Macintosh versions. In the remaining chapters, instructions have generally been combined, since the operation of many of the features is identical on both platforms. Where there are significant differences in the operation of a feature, separate Windows and Macintosh instructions are included.

Support on the Internet

Capilano Computing operates an active World Wide Web site for Logic-Works users at **http://www.capilano.com/LogicWorks**. Visit the site for program updates, installation assistance, technical support, user-contributed libraries, program add-ons and up-to-date information on using Logic-Works.

Windows Version

General Features

▓ Compatible with all current IBM PC-compatible computers running Windows 95 or newer.

▓ Fully interactive operation. Any change to a circuit, input, or device parameter immediately affects displayed circuit activity. The timing diagram is updated and scrolls continuously as the simulation progresses.

▓ LogicWorks is upward-compatible to the full DesignWorks© professional circuit-design system. All files created in LogicWorks can be read by DesignWorks. The reverse, however, is not true due to the additional structural features in DesignWorks.

Schematic Drawing Features

▓ The DevEditor module (included with LogicWorks) allows you to create libraries of custom device symbols using familiar drawing tools.

▓ Any circuit can be attached to a symbol as a subcircuit to create a simulation model. The subcircuit can be opened at any time to view or modify internal operation.

▓ A circuit schematic can be up to a total of 5 feet by 5 feet. Any number of circuit windows can be open simultaneously, allowing easy copying of partial or complete diagrams from one file to another. Each circuit is displayed in a separate window with independent control of scroll and zoom.

▓ Commands and drawing modes can be selected using menu items, keyboard equivalents, or a tool palette that is always visible in each window.

▓ Any group of objects on the drawing can be repositioned with a simple click-drag mouse action. Signal lines are rerouted interactively to maintain right-angle connections.

■ Multiple signal-line routing methods allow most pin-to-pin connections to be made with only two mouse clicks.

■ Signal names are global across a schematic. Like-named signals are thus logically connected for simulation and netlisting purposes.

■ Free text created in other programs can be pasted onto a circuit schematic. Similarly, complete or partial circuit diagrams can be pasted into word-processing or drafting documents.

■ Objects can be drawn in user-selectable colors on machines equipped with a color display.

■ Circuit and timing diagrams can be printed on any laser, inkjet, or dot-matrix printer that is supported by a Windows device driver.

Simulation Features

■ Full digital simulation capability. Circuit output may be displayed in the form of timing diagrams or on simulated output devices. Uses thirteen signal states, including forcing and resistive drive levels to correctly simulate circuits with design errors such as unconnected inputs or conflicting outputs.

■ Device delay time for individual primitive components may be set to any integer from 0 to 32,767.

■ The timing display has adjustable time-per-division and reference-line placement.

■ Common SSI and some MSI devices are implemented as primitives with hard-coded simulation functions. These can be used to create higher-level device functions. These primitive devices are "scalable," so you can create a 28-input AND gate or a 13-bit counter, for example, as a single primitive device.

■ Test and control devices, such as switches and displays, are active right on the schematic diagram, allowing circuit operation to be directly controlled and observed.

■ A Clock generator device produces signals with variable period and duty cycle. Any number of clock generators can exist in one circuit.

■ Programmable Logic Arrays can be created with up to 256 inputs and 256 outputs with user-specified binary logic. When used with ABEL$^{©}$ Student Edition Logic Compiler, PLA logic can be specified using

Boolean equations and state-transition logic. Programmable Read-Only Memories with up to 16 inputs and 128 outputs can also be simulated.

■ A simulation control palette allows the circuit to be single-stepped or run at various speeds.

■ RAM devices of any configuration from 1×1 to $1\text{Meg} \times 64$ can be created and simulated (based on available memory). Device options include 0 or 1 OE inputs, 0 to 3 CE inputs, separate- or combined-data I/O pins, and three-state or normal outputs.

New Features in Version 4

■ Completely new user interface with extensive new on-screen tools and dockable windows.

■ PLA/PROM/RAM Wizard guides you through the process of creating simulation models for these device types.

■ Add borders and title block to circuit diagrams to create finished, professional-looking printed documentation.

■ Paste graphics from any outside drawing program onto the LogicWorks schematic.

Limitations in This Version

■ The absolute maximum number of devices in a master circuit or subcircuit is 32,767.

■ A typical maximum number of devices without severe performance degradation is 500–2,000, depending on processor model.

■ The maximum length of a pin number is 4 characters.

■ The maximum length of a device, pin, or signal name is 16 characters.

■ The maximum length of a device-type name is 32 characters.

■ The maximum number of pins on a device is 800.

■ The entire circuit must fit into available memory.

Where to Start

We suggest you ease yourself into the world of schematic editing and simulation with LogicWorks by taking the following steps:

1. Install the package using the procedures outlined in Chapter 2, Getting Started, *and* read any "ReadMe" files supplied on the disk with the package.
2. If you are using LogicWorks for the first time, work first through Chapter 4, Tutorial. It provides step-by-step instructions for basic schematic editing.
3. Refer to Chapter 5, Schematic Editing, for background on basic editing techniques.

Macintosh Version

General Features

■   Fully interactive operation. Any change to a circuit, input, or device parameter immediately affects displayed circuit activity. The timing diagram is updated and scrolls continuously as the simulation progresses.

■   LogicWorks is upward-compatible to the full DesignWorks professional circuit design system. All files created in LogicWorks can be read by DesignWorks; however, the reverse is not true due to the additional structural features in DesignWorks.

Schematic Drawing Features

■   The DevEditor module (included with LogicWorks) allows you to create libraries of custom device symbols using familiar Macintosh drawing tools.

■   Any circuit to be attached to a symbol as a subcircuit to create a simulation model. The subcircuit can be opened at any time to view or modify internal operation.

▓ A circuit schematic can be up to a total of 5 feet by 5 feet. Any number of circuit windows can be open simultaneously, allowing easy copying of partial or complete diagrams from one file to another. Each circuit is displayed in a separate window with independent control of scroll and zoom.

▓ Commands and drawing modes can be selected using menu items, keyboard equivalents, or a tool palette that is always visible in each window.

▓ Any group of objects on the drawing can be repositioned with a simple click-drag mouse action. Signal lines are rerouted interactively to maintain right-angle connections.

▓ Multiple signal-line routing methods allow most pin-to-pin connections to be made with only two mouse clicks.

▓ Signal names are global across a schematic. Like-named signals are thus logically connected for simulation and netlisting purposes.

▓ Free text created in other programs can be pasted onto a circuit schematic. Similarly, complete or partial circuit diagrams can be pasted into word-processing or drafting documents.

▓ Arbitrary, user-defined text attributes can be attached to any device or signal in a circuit. This information can be used to generate SPICE-type netlists.

▓ Objects can be drawn in user-selectable colors on machines equipped with a color display.

▓ Circuit and timing diagrams can be printed on any Chooser-selectable printer, or written as PICT files for incorporation into other documentation.

Simulation Features

▓ Full digital simulation capability. Circuit output may be displayed in the form of timing diagrams or on simulated output devices. Uses thirteen signal states, including forcing and resistive drive levels to correctly simulate circuits with design errors such as unconnected inputs or conflicting outputs.

▓ Device delay time for individual primitive components may be set to any integer from 0 to 32,767.

▨ The timing display has adjustable time-per-division and reference-line placement.

▨ Common SSI and some MSI devices are implemented as primitives with hard-coded simulation functions. These can be used to create higher-level device functions. These primitive devices are "scalable," so you can create a 28-input AND gate or a 13-bit counter, for example, as a single primitive device.

▨ Test and control devices, such as switches and displays, are active right on the schematic diagram, allowing circuit operation to be directly controlled and observed.

▨ A Clock-generator device produces signals with variable period and duty cycle. Any number of clock generators can exist in one circuit.

▨ Programmable Logic Arrays can be created with up to 256 inputs and 256 outputs with user-specified binary logic. When used with MacABEL© Student Edition, PLA logic can be specified using Boolean equations and state-transition logic. Programmable Read Only Memories with up to 16 inputs and 128 outputs can also be simulated.

▨ A simulation control palette allows the circuit to be single-stepped or run at various speeds.

▨ RAM devices of any configuration from 1×1 to $1\text{Meg} \times 64$ can be created and simulated (based on available memory). Device options include 0 or 1 OE inputs, 0 to 3 CE inputs, separate- or combined-data I/O pins, and three-state or normal outputs.

New Features in Version 4

▨ PLA/PROM/RAM creation tool guides you through the process of creating simulation models for these device types.

▨ Add borders and title block to circuit diagrams to create finished, professional-looking printed documentation.

▨ Paste graphics from any outside drawing program onto the LogicWorks schematic.

Machine Compatibility

■ Compatible with all current Macintosh and Power Macintosh models with 16 megabytes of memory or more.

■ Requires MacOS version 7.5 or later; is fully System 8.0–compatible.

Limitations in This Version

■ The absolute maximum number of devices in a master or subcircuit is 32,767.

■ A typical maximum number of devices without severe performance degradation is 500–2,000, depending on Macintosh model.

■ The maximum length of a pin number is 4 characters.

■ The maximum length of a device, pin, or signal name is 16 characters.

■ The maximum length of a device type name is 32 characters.

■ The maximum number of pins on a device is 800.

■ The entire circuit must fit into available memory.

Where to Start

We suggest you ease yourself into the world of schematic editing and simulation with LogicWorks by taking the following steps:

1. Install the package using the procedures outlined in Chapter 2, Getting Started, *and* read any "ReadMe" files supplied on the disk with the package.

2. If you are using LogicWorks for the first time, work first through Chapter 4, Tutorial. It provides step-by-step instructions for basic schematic editing.

3. Refer to Chapter 5, Schematic Editing, for background on basic editing techniques.

Copyright and Trademarks

The LogicWorks software and manual are copyrighted products. The software license you have purchased entitles you to use the software on a single machine, with copies being made only for backup purposes. Any unauthorized copying of the program or documentation is subject to prosecution.

The names LogicWorks and DesignWorks are trademarks of Capilano Computing Systems Ltd. A number of other product trademarks are referred to in this manual. Full credit for these is given in the following acknowledgments.

2

Getting Started

This chapter gives you information on installing and starting up Logic-Works. The procedures for the Windows and Macintosh versions are quite different, so the chapter is divided into separate sections for these two systems. The instructions for the Windows version start below. For the Macintosh version, see page 15.

Windows Version

This section gives you information on installing and starting up LogicWorks in Windows 95, Windows 98 or Windows NT systems.

NOTE: Since the LogicWorks package and the Windows operating system are constantly being upgraded, there may be recent changes and additional information supplied in the "ReadMe" file included with the package. It is important that you review this prior to installation as it may contain information that supersedes that given here.

◆ Additional installation support is available from the LogicWorks World Wide Web site at **http://www.capilano.com/LogicWorks**.

We will be assuming that you are already familiar with general Windows operations such as copying files, creating directories, etc. If you are not, then first work through the introductory sections of the Windows user's guide.

Installation

Procedure

Follow the instructions below to install LogicWorks:

1. Start Microsoft Windows (95, 98, or NT).
2. Ensure that LogicWorks is not currently running on the target system.
3. Insert the LogicWorks CD into your CD drive.
4. On your Windows desktop, open My Computer and open the CD drive. Run the setup.exe program from the CD by double-clicking on it.
5. Follow further instructions on the screen. The installer will then unpack and copy files from the installation disk to your hard disk.
6. The installer will create a LogicWorks4 program group under Programs in the Start menu. If a "ReadMe" file's icon also appears, please read this file, as it will contain any updates to the information printed here.

General Rules

In general, you can install LogicWorks on any hard disk, in a directory of any name that you choose. Three subdirectories—"demos," "libs," and "tools"—will automatically install within that directory. If you change the names or locations of the "libs" or "tools" subdirectories, you will need to make corresponding changes to the LogicWorks initialization (.ini) file. See the following sections for more information on how to do this.

Default Library Directory

The LogicWorks initialization file allows you to name one or more directories as default library directories. All libraries in these directories will be opened automatically when the program is started. In the standard initialization file provided with LogicWorks, the default directory is called "libs."

◆ See the description of the LibraryFolder keyword in Appendix C, Initialization File Format, for more information.

Troubleshooting Installation

PROBLEM: Program seems to run, but no windows appear, and the Window and Tools menus have nothing in them.

CAUSE: The program has failed to locate any of the tool files. This could be due to any of the following:

▪ The tools are not in the tools subdirectory (normally within the same directory as the LogicWorks application), or the tools subdirectory has been renamed.

▪ The LogicWorks initialization file (lw.ini) has been renamed or is not in the same directory as the LogicWorks application.

▪ The directory name specified in the ToolFolder line of the LogicWorks initialization file is currently incorrect.

PROBLEM: Error message about "Could not open library XXXX."

CAUSE: The name or location of a library specified in the LogicWorks initialization file is incorrect.

PROBLEM: Error message about "Unexpected changes have been made to the application..." appears on startup.

CAUSE: The program checks itself for unexpected changes each time it starts up, as a precaution against viruses. If you see this message, you should run a virus-checking program on your disk and re-install the program from the master CD. In most cases, you can continue and execute the program without major ill effects, but there is a risk that you will contaminate or corrupt other files in your system.

PROBLEM: Error message about "Unable to allocate memory or bad format for module XXXX."

CAUSE: This message may appear for one of the following reasons:

▪ The module in question is obsolete; that is, it is part of an older LogicWorks release.

▪ The module file is corrupted.

▓ There is insufficient memory to load the module.

File Compatibility

This version of LogicWorks will directly read files created by earlier versions of LogicWorks, but it cannot write files compatible with these earlier versions. If you save a file using this new version, you will not be able to read it with an older version of LogicWorks.

The Initialization File

The LogicWorks initialization file (lw.ini) is a text file that specifies some initial actions that LogicWorks will take each time it is started up. The initialization file specifies the following information:

▓ Which libraries to open when the program is started.

▓ Which colors to use in displaying objects on the screen.

▓ Screen fonts, printer parameters, and other options.

The initialization file can be edited (or created anew) using your favorite programming editor or word processor. If you use a word processor, be sure to save the file using a "text only" option. The initialization file must be called "lw.ini" and must be in the same directory as the LogicWorks program itself.

◆ For more information on the meaning and format of the initialization file's contents, see Appendix C, Initialization File Format.

Macintosh Version

This section gives you information on installing and starting up LogicWorks on the Macintosh.

NOTE: Since the LogicWorks package and the Macintosh system are constantly being upgraded, there may be recent changes and additional information supplied in the "ReadMe" file included with the package. It is important that you review this prior to installation, as it may contain information that supersedes that given here.

◈ Additional installation support is available from the LogicWorks World Wide Web site at **http://www.capilano.com/LogicWorks**.

We will be assuming that you are already familiar with general Macintosh operations such as copying files, creating folders, etc. If you are not, then first work through the introductory sections of the Macintosh user's guide.

Installation

Procedure

Follow these steps to install LogicWorks:

1. Insert the LogicWorks CD into your CD drive.
2. Double-click on the "LogicWorks 4 Installer" icon.
3. Select an appropriate location on your hard disk, using the standard File Save box that the installer presents.
4. Click on the Install button.

The Installer will create a folder called "LogicWorks 4 Folder" on your hard disk and install the program files inside that folder.

NOTE: The installer will automatically choose either the PowerPC or 68K version of the package to install, based on the type of machine you run the installer on. If for any reason you wish to force it to install one or the other type, click on the Custom button on the installer screen and choose the desired version.

General Rules

In general, you can place any of the files supplied in any folder or on any disk, as suits your preference. The only absolute rule is that the LogicWorks Setup file must not be renamed and must be in the same folder as LogicWorks itself, or it will not be found at startup and no initial setup will be done. The other general warning is that if you change the names or locations of the "libs" or "tools" folders, you will need to make corresponding changes to the LogicWorks Setup file. See the following sections for more information on how to do this.

External Tools

LogicWorks consists of a small central application and a collection of external program modules that implement most of the program functions. In order for LogicWorks to recognize and load these modules, they must be either in the same folder as the LogicWorks program, or in the folder specified in the Setup file's TOOLFOLDER entry. All such tools will be loaded when LogicWorks is first started and will appear in the Tools menu. Normally, the Schematic tool will be activated at startup and will display a document window with the default filename "Design1". Other tools with more specific functions (such as PromPLA or DevEditor) are invoked by selecting their names in the Tools menu.

In this version, the only tools that are essential to normal operation of the program are Schematic and LibIO (the Parts Palette). In order to conserve memory, other tools can be kept aside in a storage folder except when actually needed.

Default Library Folder

The LogicWorks Setup file allows you to name one or more folders as default library folders. All libraries in these folders will be opened automatically when the program is started. In the standard setup file provided with Logic-Works, the default folder is called "libs."

Troubleshooting Installation

PROBLEM: Program seems to run, but no windows appear, and the Window and Tools menus have nothing in them.

CAUSE: The program has failed to locate any of the tool files. This could be due to any of the following:

- The tools are not in the same folder as the LogicWorks application, or the tools folder has been renamed.
- The LogicWorks Setup file is not in the same folder as the LogicWorks application or has been renamed.
- The folder name specified in the TOOLFOLDER line of the LogicWorks Setup file is incorrect.

PROBLEM: Error message about "Could not open library XXXX."

CAUSE: The name or location of a library specified in the LogicWorks Setup file is incorrect.

PROBLEM: Error message about "Unexpected changes have been made to the application..." appears on startup.

CAUSE: The program checks itself for unexpected changes each time it starts up, as a precaution against viruses. If you see this message, you should run a virus-checking program on your disk and re-install the program from the master diskettes (being careful to keep them write-protected). In most cases, you can continue and execute the program without major ill effects, but there is a risk that you will contaminate or corrupt other files in your system.

PROBLEM: Error message about "Unable to allocate memory or bad format for module XXXX."

CAUSE: This message can appear for one of the following reasons:

- The module in question is obsolete; that is, it is part of an older LogicWorks release.
- The module file is corrupted.

▓   There was insufficient memory to load the module.

File Compatibility

This version of LogicWorks will directly read files created using Logic-Works 3 on Macintosh or Windows. This version will *not* directly read files created by pre-3.0 versions, nor will it write files compatible with earlier versions. However, files created by all earlier versions of LogicWorks and LogiMac *can* be converted to the new format using the Converter application provided with the package. This is described in the next section.

Converting Old Files

The optional software for converting pre-3.0 files was not included in the main LogicWorks program in order to keep the size of the package down. The modules required for conversion can be installed temporarily until all your files have been converted, then removed again.

Two modules are provided:

ConvCct   Reads in any circuit file created by any previous version of LogicWorks or LogiMac, and writes out a file in version-3 format.

ConvLib   Reads in any symbol library file created by any previous version of LogicWorks or LogiMac, and writes out a file in version-3 format.

Module Installation

To install the conversion modules, do the following:

◆   If LogicWorks is already running, quit the program.

◆   Move the files ConvCct and ConvLib to the tools folder inside the LogicWorks folder.

◆   Restart LogicWorks.

The items ConvCct and ConvLib should now appear in the Tools menu.

Converting Circuit Files

The conversion operation is independent of any existing LogicWorks files that may be open. Therefore, no existing circuit files–whether open or not–will be affected by the conversion. However, in order to reduce memory requirements, it is best to perform the conversion with no other circuit files open.

To convert a file, follow the steps below:

◆ Start LogicWorks, if it is not already running.

◆ Select the ConvCct command in the Tools menu.

◆ Click the optional switches, if desired:

Convert all circuits in same folder: Checking this item will cause ConvCct to locate and convert all old-format circuit files located in the same folder as the selected files. Converted files will have the same names as the originals, plus the extension ".LW3" (or the extension specified in the Save File name). The original files are not modified or deleted unless you enter the same name for the output file.

Omit macro internals: Checking this item causes ConvCct to remove the internal logic of macro devices. This is intended for cases where the circuit is intended only for schematic purposes, and no simulation information is desired. This will substantially reduce the size of circuits that have macro components, but all simulation models will be lost.

Open converted circuit when done: If this item is checked, LogicWorks will open the converted file after conversion is complete.

◆ Click the Convert button.

◆ In the standard File Open dialog that appears, select the desired earlier-version file to be converted. If the "all circuits in same folder" option was selected, choose any circuit file in that folder.

◆ In the standard File Save dialog that appears, enter the desired name for the output file. *Do not use the same name as the input file!* If the "all circuits in same folder" option was selected, ConvCct will search for a "." in the output name you've specified, then interpret all characters following that period as a file extension, and then attach the same extension to each file converted. If no "." is found, the extension ".LW3" will be used.

When you click Save, the conversion process will start. Conversion cannot be interrupted.

NOTE: Some simulation information is not translated by ConvCct, so you may have to do a Restart or Clear Simulation to start the simulation process in the converted circuit.

Converting Library Files

The library file conversion process is very similar to the circuit conversion process described above:

◆   Start LogicWorks, if it is not already running.

◆   Select the ConvLib command in the Tools menu.

◆   Click the optional switches, if desired:

   Convert all libraries in same folder: Checking this item will cause ConvLib to convert all the old-format library files it finds in the same folder as the selected file.

   Convert subcircuits: Checking this item will cause ConvLib to produce a converted library that includes any simulation subcircuits present in the old library. You may wish to turn off this item if the symbols in the library are intended only for schematic diagram purposes and you will never want to simulate them.

◆   Click the Convert button.

◆   Using the standard file selection dialog that appears, select the library to be converted.

After you select the input file, the conversion process will start immediately. ConvLib does not prompt for an output file name. The converted library will have the same name as the input file, with the characters ".LW3" appended.

After Completing Conversion

Once all your old-format LogicWorks files have been converted, you should move the ConvCct and ConvLib modules back to the folder named "Converting Pre-3.x Files," so that they won't occupy program memory when not being used. (Only modules in the main LogicWorks folder and the tools folder are loaded when the program starts.)

Memory Usage

We recommend a memory-size setting of at least 4000K (more if you frequently work with large design files). You can set this number from the Finder by clicking once on the LogicWorks 4 program icon, then pulling down the File menu and selecting Get Info (or pressing ⌘–I).

The current release incorporates a number of new features that significantly increase the amount of memory occupied by the program itself—although the memory occupied by the circuit data has actually been reduced somewhat. Modules are loaded when the program is run and occupy memory whether they are being used or not.

The Setup File

The setup file is a text file that specifies some initial actions LogicWorks will take each time it is started up. The setup file specifies the following information:

■   Which libraries to open when the program is started.

■   Which colors to use in displaying objects.

■   Other options.

The setup file can be edited (or created anew) using your favorite programming editor or word processor. If you use a word processor, be sure to save the file using a "text only" option. The setup file must be called "LogicWorks Setup" and must be in the same folder as the LogicWorks program itself.

A typical setup file looks like the following:

LIBRARYFOLDER :Libs:;

TOOLFOLDER :Tools:;

The first word on each line is a keyword that specifies a setup option. Each statement is terminated by a semicolon (;) and can contain embedded comments in braces. The most commonly used setup items are the following:

■   LIBRARYFOLDER specifies the folder where LogicWorks will look for device library files to open.

■   TOOLFOLDER specifies the folder where LogicWorks will look for code module files.

3

User Interface

This chapter provides general information on the use of windows, drawing tools, and other user interface features of LogicWorks. Since these features are substantially different in the Windows and Macintosh versions of the program, this chapter is divided into separate sections for each of these systems. The Windows version is described starting below; for the Macintosh version see page 34.

NOTE: Some minor differences exist between the Windows and Macintosh system in the terminology used to described user interface elements and system features. For example "directory" vs. "folder," "toolbar" vs. "tool palette," and "Finder" vs. "Explorer." Rather than complicating the manual with constant splits for the two systems, we will use terminology from one or the other system where it is close enough to be understandable. Our apologies to platform purists!

Windows Version

Mouse Button Usage

Three different mouse button actions are used for various functions in LogicWorks. For clarity, we will use the following terminology when referring to these actions in the remainder of the manual:

Click   means press and release the left button without moving the mouse. Example: To select a device, click on it.

Click and drag   means press the left mouse button and hold it down while moving the mouse to the appropriate position for the next action. Example: To move a device, click and drag it to the desired new position.

Double-click   means press and release the left button twice in quick succession without moving the mouse. Example: To open a device's internal circuit, double-click on the device.

Right-click   means press the right button. This is usually used to display a pop-up "context" menu that allows you to perform operations on the selected item.

Dialog Boxes

Many LogicWorks functions display information or request information by displaying a window called a dialog box. For example, the following dialog box is displayed when a Get Info command is executed for a signal.

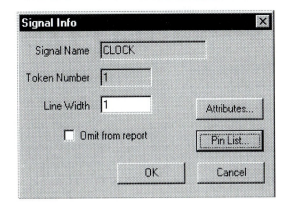

Enter Key

In general, the controls in dialog boxes will behave according to Windows standards, with the following exception: Since some of the boxes require text entry with multiple lines, the (ENTER) key *will not* cause the box to be closed and the default action to be performed in these cases. If the flashing text cursor is currently in a text box that allows multiple line entry, such as the Attributes Dialog, then the (ENTER) key will simply insert a hard return in the text. If the insertion point is located in a text box that does *not* allow multiple lines, then the (ENTER) key will execute the dialog box's default action.

Using the Clipboard in Dialog Boxes

In most dialog boxes requiring text entry, the keyboard equivalents for the Clipboard commands Cut ((CTRL)–X), Copy ((CTRL)–C), and Paste ((CTRL)–V) are active and can be used to transfer text to or from a text box.

Window Usage

Circuit Windows

Each circuit window displays a circuit schematic. The title on a schematic window will be the name of the circuit file displayed in that window, as in: DESIGN9.CCT. Any number of circuits can be displayed simultaneously. At any given time, only one circuit is "current"—the one in the topmost window. Any other window can be made current simply by clicking the mouse anywhere in that window.

Timing Windows

When you are simulating a circuit, you can optionally display a Timing diagram window to display signal values versus time. Only one Timing window can be displayed at any given time, and it shows waveforms generated by the current design. Closing the Timing window does not close the circuit design file.

Window Operations

Circuit and Symbol Editor Windows

This table summarizes the common operations on circuit and symbol editor windows.

To:	Do this:
Close the window	Click on the Close icon (the X at the top right corner of the window) closes the window. If the circuit being closed was the top level of the design (i.e., it wasn't an internal circuit of a device symbol) then the circuit file is closed.

Resize the window

To enlarge or reduce a circuit or symbol editor window, position the mouse pointer along any edge or corner of the window, and click and drag the window to the desired size. As long as the mouse button is held down, a gray outline of the window will track the mouse movements. When the button is released, the window will be redrawn to the new size and shape.

To expand the circuit window to fill the whole available area, click on the Maximize button at the upper right corner of the window. Clicking again on this button will cause the window to return to its original size.

Move the window

To move the circuit or symbol editor window, position the mouse pointer on the title bar along the top edge of the window and press and hold the mouse button. As long as the button is held down, a gray outline of the window will track the mouse movements. When the button is released, the window will be redrawn at its new position.

Timing, Parts and Other Docking Windows

The timing window, parts palette and other display and control windows are referred to as "docking" windows. This means that they can be positioned at one edge of the main application window and they will automatically adjust themselves to fit as the main application is moved and resized. This table summarizes the positioning controls for docking windows.

To: Do this:

Close the window

Click on the Close icon (the X at the top right corner of the window) closes the window.

Reopen a closed window

Docking windows are normally associated with a specific program function and may not be displayed at all times. The timing window, parts palette and other commonly-used windows have a button associated with them on the toolbar. Clicking this button will show/hide the associated window.

Resize the window

Docking windows can be moved to any edge of the main application window by clicking and dragging in the "gripper bar" of the window. If a docking window is dragged outside the main application, it will "float" as a separate window.

Float the
window
Docking windows can be switched to "floating" windows, that is, a separate window that stays above all other windows and can be resized independently. To float a window, right-click in the top bar (near the close box) of the window and uncheck the "Allow Docking" menu item. You can now move the window to any desired location within or outside the main application window.

The Window Menu

Selections provided on the Window menu (on the LogicWorks menu bar) can be used to bring to the front any window that is currently open.

Keyboard Usage

The keyboard is absolutely required only when entering names for devices or signals, or for placing free text notations on the drawing. However, the (CTRL), (SHIFT), and (TAB) keys on the keyboard can be used with many editing operations to invoke special features such as auto-alignment, auto-naming, and different signal-line drawing methods. In addition, the "arrow" keys can be used as a convenient way of setting symbol rotation while placing devices or pasting circuit groups. These options are described in detail in the relevant chapters.

Pop-Up Menus

At any time while editing a diagram, you can use the right mouse button to click on a schematic object. A pop-up menu will appear under the cursor, allowing you to select from commands appropriate to that object. For example, the menu for a device contains commands to get device information, edit attributes, open the internal circuit, flip or rotate the symbol, perform Cut and Copy operations, and so on.

Separate pop-up menus are available for devices, signals, pins, attributes, and (if you click on open space on the drawing) the circuit itself.

◆   For details on commands in pop-up menus, see Chapter 12, Menu Reference.

Toolbars

Toolbars are collections of graphical buttons that remained anchored to the edge of the main program window. They provide direct access to many program functions as an alternative to menu selections. Here is a typical example:

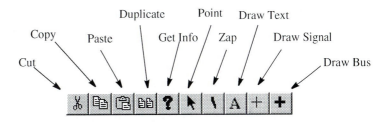

Learning the Tool Functions

Several on-screen hints help you determine the function of a tool that may not be familiar to you. As you move the cursor over a tool, a description of the tool appears in the status bar at the bottom of the screen. In addition, if you pause over a tool, a small window will pop up momentarily with the title of that tool.

Moving the Toolbars

Toolbars can be moved to any desired position on the screen by clicking and dragging anywhere in the toolbar that is not on a button. If the toolbar is moved away from the top of the main application window it will become a floating palette that will stay in front of all other windows.

Hiding a Toolbar

A toolbar can be hidden by either of these methods:

▦ Drag the toolbar away from the top of the main application window so that it becomes a floating palette, then click on its close ("X") icon in the top right corner.

▦ Select the associated menu item in the View menu. The menu item will be checked when the toolbar is showing. Selecting the item will uncheck it and hide the toolbar.

Use of the Pointer or Cursor

In subsequent descriptions, we will refer to the small onscreen shape that tracks the mouse position as the "pointer" or "cursor," In LogicWorks there are a number of cursor modes which determine what action will be performed when the mouse button is clicked. Following is a summary of cursor modes. More detailed descriptions of operations performed in each mode are provided in later reference chapters.

Note that the cursor shape sometimes differs from the Toolbar icon for ease of pointing.

Tool Palette Icon	Initial Cursor Shape	Equivalent Menu Command	Description
▶	▶	Point	Used to select or drag objects, extend signals.
⌇	⌇	Zap	Used to remove single objects. Press the button to remove whatever the tip of the cursor is pointing at. Objects can also be removed in groups by selecting them and using the Clear command or Delete key.

T		Text	Used to select a signal or device to name, or to place free text on the diagram. Point at the item you want to name and press and hold the mouse button. Move to where you want the name to appear, then release the button.
+		Draw Bus	Used to create a new bus line or extend an existing bus. Clicking once fixes a corner, double-clicking terminates the line.
+	×	Draw Sig	Used to create a new signal line or extend an existing signal. Clicking once fixes a corner, double-clicking terminates the line. Note that most signal-drawing operations can also be done in Point mode.
		Magnify	Used to zoom in and out. Clicking on a point, or dragging down and right, zooms in. Dragging up and left zooms out.
		Signal Probe	Used to view and change signal values on the schematic. Click on the signal to view with the end of the probe and the cursor follows changes in signal value. Type "0" or "1" on the keyboard to enter a value. See Chapter 7, Simulation, for more information.

Parts Palette

Parts library contents are displayed in a floating palette window which looks like the following:

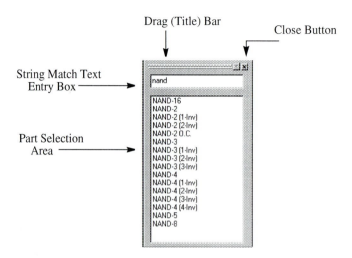

The Parts Palette displays the contents of all parts available in all open library files and allows any item to be selected for placement within the schematic.

Moving the Parts Palette

The Parts Palette can be moved to any desired location on the screen by clicking and dragging its title bar.

Hiding the Parts Palette

The Parts Palette can be removed from the screen by clicking in its close button. To re-display the palette, select the Parts Palette command from the View menu.

Choosing a Library

The palette displays the contents of all open library files, merged into a single list. Library files are opened by selecting the Open command in the Parts Palette pop-up menu, discussed in a following section.

Selecting a Part

To select a part for placement in the schematic:

◆ If necessary, use the scroll bar to scroll the list until the desired part name is in view.

◆ Double-click on the part name in the list.

◆ Move the cursor to the current schematic window to place the part.

Using the String-Matching Box

The string match text box allows you to type characters that will reduce the size of the list and make it easier to locate the desired part. Simply click in the text box and type the desired characters. After a brief pause, the displayed part list will be reduced to only those parts that contain the typed string of characters.

To return to the full selection, click at the end of the characters in the box and backspace over them until the box is cleared.

Parts Palette Pop-Up Menu

The Parts Palette pop-up menu can be activated by pressing the right mouse button in this window. This menu allows you to open, close, and manage your libraries.

◆ See Chapter 12, Menu Reference, for specific information on these items.

Macintosh Version

This chapter provides general information on the use of windows, drawing tools, and other user interface features of LogicWorks.

Mouse Button Usage

Three different mouse button actions are used for various functions in LogicWorks. For clarity, we will use the following terminology when referring to these actions in the remainder of the manual:

Click

means press and release the mouse button without moving the mouse. Example: To select a device, click on it.

Click and drag

means press the button and hold it down while moving the mouse to the appropriate position for the next action. Example: To move a device, click and drag it to the desired new position.

Double-click

means press and release the button twice in quick succession without moving the mouse. Example: To open a device's internal circuit, double-click on the device.

Dialog Boxes

Many LogicWorks functions display information to—or request information from—the user. They do so by displaying a window called a dialog box. For example, the following dialog box is displayed when a Get Info command is executed for a signal.

Enter vs. Return Keys

In general, the controls in dialog boxes will behave according to Macintosh standards, with the following exception: Since many of the boxes require text entry with multiple lines, the Return key *does not* normally cause the box to be closed and the default action to be performed; this function has been given to the (ENTER) key. If the flashing text cursor is currently in a text box that allows multiple line entry, such as the attributes box, then the Return key will simply insert a carriage return in the text. If the insertion point is located in a text box that does *not* allow multiple lines, then the Return key will behave the same as the Tab key and cause the next text box to be activated.

Using the Clipboard in Dialog Boxes

In most dialog boxes requiring text entry, the keyboard equivalents for the Clipboard commands Cut (⌘–X), Copy (⌘–C), and Paste (⌘–V) are active and can be used to transfer text to or from a text box.

Window Usage

Circuit Windows

Each circuit window displays a circuit schematic. The title on a circuit window will be the name of the circuit file displayed in that window, as in "Interface Board." Any number of circuits can be displayed simultaneously. At any given time, only one circuit is "current"—the one in the topmost window. Any other window can be made current simply by clicking the mouse anywhere in that window.

Timing Windows

When you are simulating a circuit, you can optionally display a Timing diagram window to display signal values versus time. One Timing window can be displayed per design that is open. Closing the Timing window does not close the design file.

Closing a Window

Clicking in a circuit window's close box has the effect of closing the circuit. If the circuit being closed was the top level of the design (i.e., it wasn't an internal circuit of a device symbol) then the circuit file is closed.

Resizing a Window

To enlarge or reduce a circuit or Timing window, position the pointer in the size box at the lower right corner of the window and press and hold the mouse button. As long as the mouse button is held down, a gray outline of the window will track the mouse's movements. When the button is released, the window will be redrawn to the new size and shape.

To expand the circuit window to fill the whole screen, click on the zoom box at the upper right corner of the window. Clicking again in this box will cause the window to return to its original size.

Moving a Window

To move the circuit or Timing window, position the pointer in the title bar anywhere along the top edge of the window and press and hold the mouse button. As long as the button is held down, a gray outline of the window will track the mouse movements. When the button is released, the window will be redrawn at its new position.

The Window Menu

Selections provided on the Window menu can be used to bring to the front any window that is currently open.

Keyboard Usage

The Macintosh keyboard is absolutely required only when entering names for devices or signals, or for placing free text notations on the drawing. However, the ⌘, (SHIFT), and (OPTION) keys on the keyboard can be used with many editing operations to invoke special features such as auto-alignment, auto-naming, and different signal-line drawing methods. In addition, the "arrow" keys (if available on your keyboard) can be used as a convenient way of setting symbol rotation while placing devices or pasting circuit groups. These options are described in detail in the relevant chapters.

Pop-Up Menus

At any time while editing a diagram, you can hold the Command key and click on a schematic object. A pop-up menu will appear under the cursor, allowing you to select from commands appropriate to that object. For example, the menu for a device contains commands to get device information, edit attributes, open the internal circuit, flip or rotate the symbol, perform Cut and Copy operations, etc.

Separate pop-up menus are available for devices, signals, pins, attributes, and (if you click on open space on the drawing) the circuit itself.

◆ For details on commands in pop-up menus see Chapter 12, Menu Reference.

Tool Palette and Status Display

The Tool Palette is a small window which always remains in front of any circuit windows currently open. The individual items in this display are described in following sections.

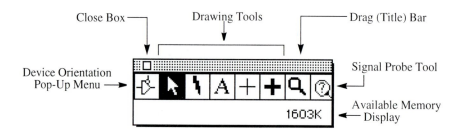

Moving the Tool Palette

The Tool Palette can be moved to any desired position on the screen by clicking and dragging its title bar.

Hiding the Tool Palette

The Tool Palette can be removed from the screen by clicking once on its close box. To re-display the palette, select the Show Tool Palette menu item in the Edit menu.

Available Memory Display

If this option is enabled, the amount of free memory available for circuit data is displayed in the lower right corner of the Tool Palette window. This number is shown in Kbytes, i.e. multiples of 1024 bytes. The available memory display option can be enabled or disabled through the Design Preferences command in the Schematic menu.

Orientation Control

The device orientation icon shows the orientation that will be used when the next device or circuit scrap is pasted into the schematic. The orientation can be changed by clicking on the icon, by using the arrow keys on the keyboard, or by using the Orientation menu command.

Clicking and holding the device orientation icon in the Tool Palette will cause the following pop-up menu to appear:

Choose any one of the eight possible orientations by sliding the pointer across this menu and releasing the mouse button when the orientation you want is highlighted. The selected icon orientation will be displayed in the Tool Palette icon.

This selection has no effect on items already placed in the schematic. It affects only the flickering image of items currently being placed and devices selected later from the Parts Palette.

◈ More information on symbol rotation is given in Chapter 6, Advanced Schematic Editing.

Schematic Drawing Tools

The seven drawing tools are used to set the cursor mode for schematic editing. More information on cursor modes is given in the following section.

Use of the Pointer or Cursor

In subsequent descriptions, we will refer to the small shape that tracks the mouse position on the screen as the pointer or cursor. In LogicWorks there are a number of different cursor modes used which determine what action will be performed when the mouse button is clicked. Following is a summary of the cursor modes. More detailed descriptions of operations performed in each mode are provided in later reference chapters.

Note that the cursor shape sometimes differs from the Tool Palette icon for ease of pointing.

Tool Palette Icon	Initial Cursor Shape	Equivalent Menu Command	Description
▲	▲	Point	Used to select or drag objects, extend signals.
⚡	⚡	Zap	Used to remove single objects. Press the button to remove whatever the tip of the cursor is pointing at. Objects can also be removed in groups by selecting them and using the Clear command or Delete key.
A	⟋	Text	Used to select a signal or device to name, or to place free text on the diagram. Point at the item you want to name and press and hold the mouse button. Move to where you want the name to appear, then release the button.
+	×	Draw Sig	Used to create a new signal line or extend an existing signal. Clicking once fixes a corner, double-clicking terminates the line. Note that most signal-drawing operations can also be done in Point mode.

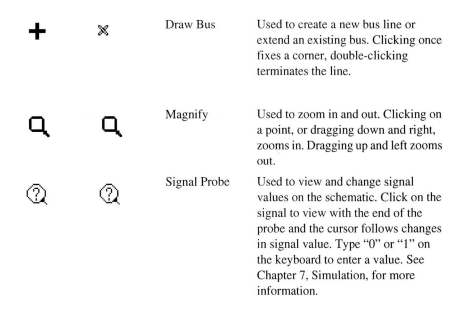

		Draw Bus	Used to create a new bus line or extend an existing bus. Clicking once fixes a corner, double-clicking terminates the line.
		Magnify	Used to zoom in and out. Clicking on a point, or dragging down and right, zooms in. Dragging up and left zooms out.
		Signal Probe	Used to view and change signal values on the schematic. Click on the signal to view with the end of the probe and the cursor follows changes in signal value. Type "0" or "1" on the keyboard to enter a value. See Chapter 7, Simulation, for more information.

Parts Palette

Parts library contents are displayed in a floating palette window which looks like the following:

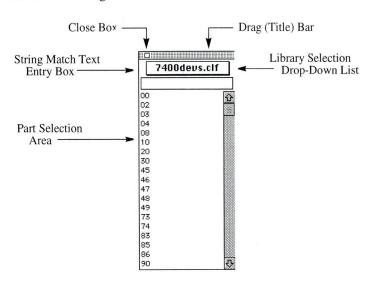

The Parts Palette displays the contents of the selected library file and allows any item to be selected for placement in the schematic.

Moving the Parts Palette

The Parts Palette can be moved to any desired location on the screen by clicking and dragging in its drag bar.

Hiding the Parts Palette

The Parts palette can be removed from the screen by clicking on its close box. To re-display the palette, select the Show Part Palette command in the Libraries submenu folder of the File menu.

Choosing a Library

The palette displays the contents of only one library at a time. The library currently being displayed is shown in the library selection box at the top of the palette. Clicking in this drop-down list box allows you to select from among any of the open library files. Library files are opened by selecting the Open Lib command, either in the Libraries submenu of the File menu or in the Parts Palette pop-up menu discussed below.

Selecting a Part

To select a part for placement in the schematic:

1. If necessary, use the scroll bar to scroll the list until the desired part name is in view.
2. Double-click on the part name in the list.
3. Move the cursor to the current schematic window.

Using the String-Matching Box

The string match text box allows you to type characters that will reduce the size of the list and make it easier to locate the desired part. Simply click in the text box and type the desired characters. After a brief pause, the displayed part list will be reduced to only those parts that contain the typed string of characters. For example:

To return to the full selection, click at the end of the characters in the box and backspace over them until the box is cleared.

Parts Palette Pop-Up Menu

As a shortcut, the Libraries submenu in the File menu can be popped up right in the Parts Palette. Hold down the Command key and, with the mouse pointer anywhere in the Parts Palette, press and hold the mouse button. Selections that you make from the resulting pop-up menu are equivalent to selecting the same items in the Libraries submenu.

◆ See Chapter 12, Menu Reference, for specific information on these items.

4
Tutorial

This tutorial is divided into a number of sections, allowing you to review the basic functions first, then learn about more advanced features. The first section is entitled "The Five-Minute Schematic and Simulation" and will give you a taste of how quickly you can put together a circuit with full simulation. The later sections are divided by subject, so you can study in greater detail the features that are important for your application.

These tutorials are intended only to introduce you to LogicWorks features. For complete details on any subject, see the reference sections of this manual.

Tutorial Manual Format

In the following tutorial sections, text with a diamond:

◆ like this

provides step-by-step instructions for achieving a specific goal. Other text provides background and explanation of the actions being taken.

The Five-Minute Schematic and Simulation

In this section, we're going to show how quickly you can create and test a circuit using LogicWorks.

Starting LogicWorks

◆ Start the LogicWorks program by double-clicking on its icon.

Once the program has started, you will be looking at a screen like this.

Windows Version

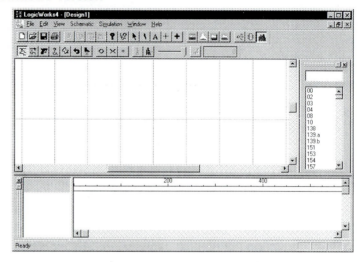

Macintosh Version

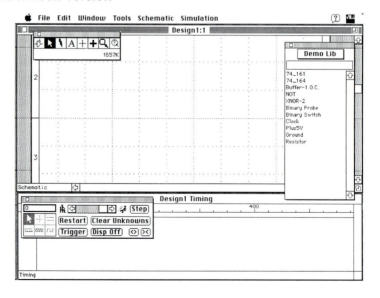

The "Design" window is your viewport onto the circuit diagram, which you will manipulate using the various drawing tools. The smaller "Timing" window will be used by the program to display a timing diagram of the signals in your circuit. Either of these windows can be moved or resized by the usual methods, to suit your needs.

Placing a Device

Windows—The parts palette shows a merged list of all parts in all open libraries. Libraries can be opened and closed manually using the Parts pop-up menu's Open and Close commands, or any collection of libraries can be opened automatically at startup by placing them in the "Libs" directory.

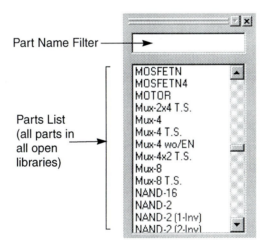

Part Name Filter

Parts List
(all parts in
all open
libraries)

MOSFETN
MOSFETN4
MOTOR
Mux-2x4 T.S.
Mux-4
Mux-4 T.S.
Mux-4 wo/EN
Mux-4x2 T.S.
Mux-8
Mux-8 T.S.
NAND-16
NAND-2
NAND-2 (1-Inv)
NAND-2 (2-Inv)

Macintosh—The parts palette shows the parts from a single library at any one time. To select which library you are viewing, click on the library selector pop-up at the top of the palette. For this example, select the library "Demo Lib."

The list of open libraries can be changed manually by doing either of the following:

■ Pull down the File menu on the Menu Bar and highlight Libraries; then slide right to the submenu containing New Lib, Open Lib, and so on, and select the appropriate command.

■ ⌘-click on the Parts Palette; then select from a pop-up menu that contains the first five of the six options that are listed on the submenu.

Any collection of libraries can be opened automatically at startup by modifying the Setup file.

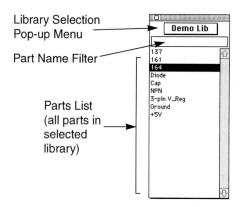

Library Selection
Pop-up Menu

Part Name Filter

Parts List
(all parts in
selected
library)

◆ Locate the part 74_164 in the parts list and double-click on it.

◆ Move the cursor back into the circuit window. The cursor on the screen will now be replaced by a moving image of the selected symbol, in this case an 8-bit shift register.

The numbered devices in this library are generic 7400-series types. The labeling and simulation characteristics can be adjusted to match the various 7400 families on the market.

◆ Position this image somewhere near the center of the circuit window and click the mouse button. A permanent image of the device will now stay behind in that location and the image will continue to follow your movements.

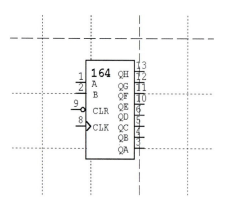

More devices of the same type could be created at this point, but in this example we wish to select another symbol.

◆ Press the spacebar to return to Point mode. Notice that you can click and drag the device that you placed to any desired new position.

◆ Move again to the Parts Palette and this time double-click on the XNOR-2 type. Once you move outside of the Parts Palette, the cursor will immediately change to match the new symbol.

The XNOR-2, and the devices in the primgate.clf, primlogi.clf, and primio.clf libraries, are called "primitive" types because they have built-in simulation models in LogicWorks. Other devices, such as those in the 7400devs.clf library are called "subcircuit" types because their simulation models are made up of primitives. If LogicWorks is being used only for schematic entry, it is also possible to make symbols with no simulation function.

◆ Place one of these Exclusive-NOR gates adjacent to the 164 device so that the pins just touch, and click once to anchor the device.

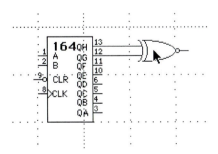

◆ Press the spacebar to return to Point mode.

Whenever you place devices or signal lines so that they touch, you will notice that the signal lines flash briefly. This indicates that a logical connection has been made. You do not need to explicitly request a connection.

Moving a Device

◆ Point at the Exclusive NOR gate and click and drag to the right. While you hold the mouse button you can drag the device to any desired new position. Note that any signal lines attached to the device are adjusted continuously to maintain connection.

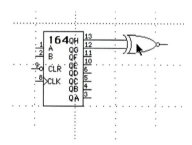

◆ Position the gate as shown to the right of the 164 device.

Drawing Signal Connections

◆ Attach a connection to the output of the gate by positioning the pointer near the endpoint of the pin and dragging away to the upper left.

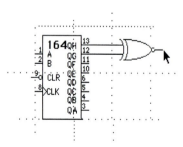

◆ Notice that two lines at right angles will follow your mouse movements to connect the starting and ending points.

Windows—While moving the mouse, try pressing the (CTRL) and/or (TAB) keys and note the different line-routing methods available. Click mouse once to anchor the signal line.

Macintosh—While moving the mouse, try pressing the ⌘ and/or (OPTION) keys and note the different line-routing methods available.

◆ For details on these line-routing modifier keys, see the section on Signal Line Editing in Chapter 5, Schematic Editing.

◆ Leave a right-angle line attached to the gate, as shown.

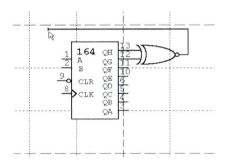

◆ Extend this line to connect to the B input of the 164 by clicking at the line endpoint where you left off, dragging the line to the B input, and releasing the mouse button.

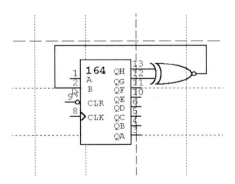

◆ Add a connection to pin A by clicking at the end of the pin, dragging the line down until it touches the signal line, then releasing the mouse button.

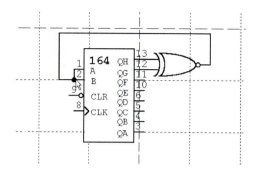

Notice that an intersection dot appears automatically whenever three or more lines intersect.

◆ Try repositioning a line segment by clicking and dragging anywhere along the length of the segment except at a corner or intersection.

Binary Switch Input Device

◆ Return to the Parts Palette and select a Binary Switch device from the demolib.clf library.

◆ Place it as shown on the diagram.

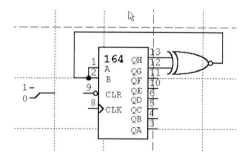

◆ Press the spacebar to return to Point mode.

◆ Try clicking on the switch. Notice that it changes between the 0 and 1 states.

In order to move a switch, you must first select it by holding the (SHIFT) key while clicking on it. This is necessary because the switch has a special response to a normal mouse click.

The devices in the Simulation IO library can be used to actively control and observe the simulation right on the schematic. Each of these devices responds immediately to changes in the simulation in progress. The Hex Keyboard device is similar to the switch except that it operates on four lines at once.

Clock Generator Device

◆ Select a Clock device from the demolib.clf library and place it on the diagram just below the switch.

◆ Press the spacebar to return to Point mode.

◆ Route wires from the switch and clock to the 164, as shown. Remember to try using the (CTRL) and (TAB) keys on the Windows keyboard, or the (OPTION) and ⌘ keys on the Macintosh keyboard, to route the wires.

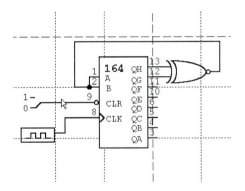

While you have been working on the diagram, the LogicWorks simulator has been running continuously, simulating the effects of the new connections that are being made. So far, though, we have not asked it to display any results. This is done either by placing probes on the diagram or by displaying signals in the Timing window.

Naming a Signal

◆ Click on the text icon in the Tool Palette. The cursor will then change to a pencil shape, which will be used to select the item we want to name.

Text Tool

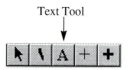

The text cursor is used to name devices and signals, to apply pin numbers
to device pins, and to add free text notations to the diagram.

◆ Position the tip of the pencil anywhere along the length of the line
running from the clock device, and press *and hold* the mouse button.
The cursor will change to an I-beam shape.

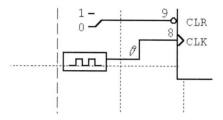

◆ Still holding the mouse button pressed, move the cursor down
somewhere below the signal line.

◆ Release the mouse button. A blinking insertion marker will appear.

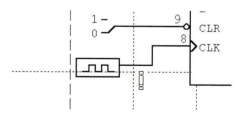

◆ Type the name "CLK" on the keyboard, then press the (ENTER) key or
click the mouse button once.

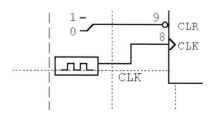

◆ Return to Point mode by clicking the arrow icon in the Tool Palette. Note that the name can be dragged to any desired position.

◆ Click once on the Binary Switch to change it to the logical 1 state.

The Timing Window

You will immediately see the Timing window come to life with the displayed values on the CLK line. By default, any named signal is shown automatically in the Timing window.

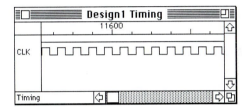

You can disable this feature, if desired:

Windows—Pull down the Simulation menu and deselect the Add Automatically command.

Macintosh—Pull down the Simulate menu and deselect the Auto Add command.

◆ Again using the text cursor, name the two data lines from the shift register and the output line from the gate, as shown. The simulated output from these lines will immediately appear in the Timing window.

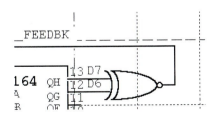

Simulation Controls

Windows—Click on the <> and >< buttons and observe that they affect the time scale of the Timing window.

Macintosh—

◆ Select Sim from the Tools menu to make the Simulator Palette visible.

◆ Click on the >< and <> buttons and observe that they affect the time scale of the Timing window.

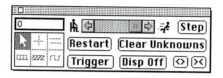

Display resolution can be adjusted from 4 pixels/time unit to 100 time units/pixel. The interpretation of a "time unit" is arbitrary, but it is convenient to think of it as a nanosecond.

Windows—

◆ Select the Timing Window item in the View menu. You will notice that the Timing window disappears and the current time indicator in the Simulator palette advances much more quickly.

◆ Select the Timing Window item again to re-enable the display.

◆ Click on the Reset (⤺) button and notice that the simulation restarts at time 0.

Macintosh—

◆ Try clicking on the Disp Off button. You will notice that the Timing window goes blank, and the current time indicator in the control palette advances much more quickly.

◆ Click again on the same button—which now reads "Disp On"—to re-enable the display.

◆ Click on the Restart button and notice that the simulation restarts at time 0.

Windows—Adjust the speed slider control in the Simulator toolbar and notice that simulation slows.

◆ Click repeatedly on the Step (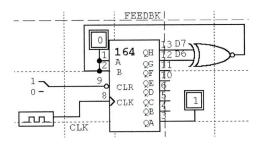) button and observe that the simulation proceeds one step at a time.

◆ Click the Run button in the Simulator toolbar.

Macintosh—On the Simulator Control Palette, click on the arrow at the left end of the speed control bar. Notice that each click slows the simulation by one step. Continue clicking until the simulation stops.

◆ Click on the Step button and observe that the simulation proceeds one step at a time.

NOTE: The Step button advances the simulation to the next time at which there is some circuit activity, not necessarily just one time unit. The size of the step will depend upon the circuit.

Probe Device

◆ Select the Binary Probe type from the Parts Palette.

◆ Place a probe so that its pin contacts a signal line to view the simulation value on that line.

As the simulation progresses, the values on all probes are updated immediately. A similar device, the Hex Display, is also available to show groups of lines in hexadecimal. These simulation devices can be flagged to indicate that they are not a real part of the finished product and should not be included in any netlists or bills of materials.

Setting Device Parameters

◆ Click in the window, but away from any circuit objects. This deselects everything.

◆ Click on the XNOR gate to select it.

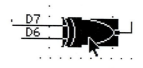

Windows—Select the Simulation Params command in the Simulation menu.

Macintosh—Select the Simulation Params command in the Simulate menu.

◆ Click on the "+" button a couple of times to increase the propagation delay in this device.

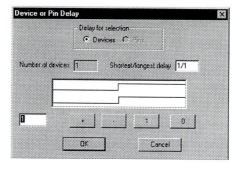

The Simulation Params command is used to view and set delays associated with devices and pins. Pin delays normally default to zero but can be used to fine-tune the delays for different paths through a device.

◆ Click on the OK button.

Device Delay on the Timing Window

Notice that the altered device delay immediately affects the simulation. You will see an increased delay between the clock reference lines and the changes in the FEEDBK signal.

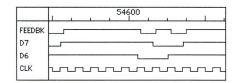

Interacting with the Simulation

◆ Try clicking on the switch hooked to the CLR input. Notice that it changes state and immediately affects the displayed simulation results.

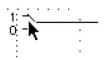

Saving the Design

Windows—Click the Save button (), and save your circuit so you can continue with it later.

Macintosh—Select the Save Design As command in the File menu and save your circuit so you can continue with it later.

This ends the Five-Minute Schematic and Simulation tutorial section.

Schematic Editing

The object of this tutorial is to take a closer look at LogicWorks' schematic editing features. We will do this by making a number of modifications to the circuit file created in the Five-Minute Schematic and Simulation demonstration.

Topics covered in this section:

- ▓ Deleting and moving objects
- ▓ Selecting device types by name
- ▓ Device symbol rotation
- ▓ Using power and ground connectors
- ▓ Connecting signals by name
- ▓ Using Copy and Paste on circuit objects
- ▓ Naming devices
- ▓ Adding pin numbers to devices
- ▓ Placing text notations on the diagram.

Opening a Circuit File

◆ Open the file you created in the initial demonstration.

Windows—Click on the Open button ( ) or select the Open... item in the File menu.

Macintosh—Select the Open Design command on the File menu.

The file created in the first section is also supplied with LogicWorks in the demos directory/folder.

Deleting Objects

◆ Mac only: if the Tool Palette is not visible, click once in the circuit window.

◆ Select the Zap (lightning bolt) tool in the Tool Palette. The cursor will change to match this icon.

This tool is used to remove a single object from the diagram. When aimed at a device, the device is removed. When aimed at a signal line, the line segment is removed to the nearest device pin or intersection.

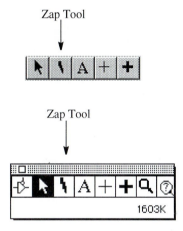

Zap Tool

Zap Tool

Breaking a Signal Line

◆ Zap the segment shown.

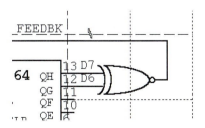

The signal has now been broken into two pieces and the signal name FEEDBK will become associated with the closest piece. You can click on the name to see which piece highlights.

If the simulation is running, notice that the signal values are updated to reflect the unknown input values on the shift register.

Rotating a Device

◆ In the Parts Palette, double-click on the device "O.C. Buf -1" (in the demolib.clf library). Move the mouse pointer over to the Schematic window.

◆ Before placing the device, try pressing the arrow keys on the keyboard. Notice that this changes the orientation of the device symbol that is moving on the screen. You can also click on the device orientation icon in the Tool Palette to change its orientation.

	Press left arrow key for	
	Press it again for	
	Press right arrow key for	
	Press it again for	
	Press down arrow key for	
	Press it again for	
	Press up arrow key for	
	Press it again for	

There are actually eight different orientations available: the four major compass points, plus these directions with an additional flip around the major axis. The device orientation icon in the Tool Palette shows the current selected orientation. This orientation also applies when groups of objects are pasted or duplicated.

Placing a Device

◆ Place one of these open-collector buffers on the diagram as shown.

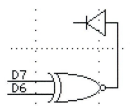

The open-collector buffer is a device that converts a low input into a low-impedance output, and a high input to a high-impedance output. How this works in conjunction with a pullup resistor will be demonstrated in the following sections.

Resistor Device

◆ In the Parts Palette, select a Resistor device from the demolib.clf library, orient it, and place it as shown.

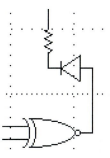

The Resistor device has special properties in the digital simulation. It conducts signal values in both directions but its output has a lower drive level than its input. Thus, it can be used as a pullup or pulldown resistor in circuit logic, or as a series resistor to simulate low-drive-level devices.

Power and Ground Connectors

◆ Select a +5V device from the Parts Palette and place it.

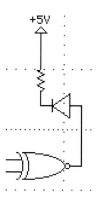

The +5V symbol is called a Signal Connector and performs several functions. It puts out a constant high level for use in the simulation, it assigns the name "Plus5V" to the attached signal line, and it creates a logical connection to all other signal lines having the same symbol attached. You can create your own Signal Connectors for commonly used signals using the DevEditor module.

Clearing the Simulation

◆ Press the spacebar to reactivate the arrow pointer, then wire these devices together as shown.

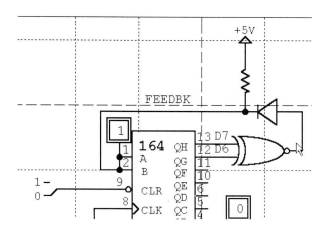

The Timing window and probes will now be showing unknown values due to the broken connections and feedback in the circuit. You can clear these by selecting the Clear Unknowns command from the Simulation menu, or by clicking on the Clear button in the Simulator tools.

Dragging Groups of Devices

◈ Select the three symbols shown by (SHIFT)–clicking on them. Then drag them as a group to a more centered position, as shown.

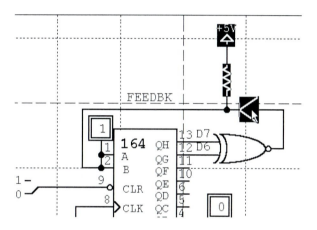

Connecting Signals by Name

◈ Select a NOT (Inverter) from the Parts Palette and place below the 164 device, as shown.

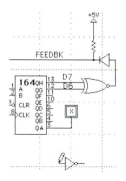

◆ Select the Text cursor and name the input on the NOT device to D7, as described in the introductory demo. Be careful to click the pencil at the *end* of the pin. Clicking in the middle of the pin will create a pin number.

When you place the label D7, you will notice this line will flash, indicating that a logical connection has been made. For simulation and netlisting purposes, these two signals are now connected together. Any like-named signals are considered to be connected.

◆ Double-click on either of the D7 signal lines to check connectivity (both lines will change color).

◆ Label the output of the inverter NOTD7 and observe the inverted signal in the Timing window.

Using Copy and Paste

◆ To copy a section of the circuit to a new circuit file:

◆ Select the group of circuitry to be copied by clicking and dragging across a group of objects. Any object that intersects the rectangle will be selected.

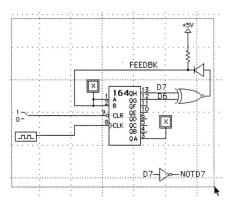

◆ Select the Copy command from the Edit menu.

◆ Select the New Design (Mac) or New (Windows) command from the File menu and create a new, empty design.

◆ Select the Paste command in the Edit menu.

An outline of the entire circuit will now follow your mouse movements. You can place this group of objects anywhere on the new diagram. You can also use the arrow keys to reorient the circuit group before placing it.

The Cut, Copy, Paste, and Duplicate commands can be used on any single object or any group of selected objects.

◆ Select Close Design on the File Menu to close the extra copy of the circuit without saving. The original circuit window should still be open.

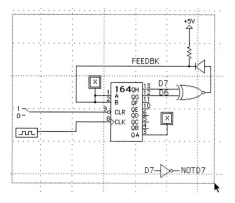

Naming Devices

Devices are named in a manner similar to signals:

◆ Select the text cursor in the Tool Palette.

◆ With the pencil inside the device, click the mouse button and hold it pressed.

◆ Move to the desired position for the name.

◆ Release the mouse button.

◆ Type the desired name.

◆ Press the (ENTER) key or click the mouse button once.

◆ Repeat the above steps to name the devices as shown.

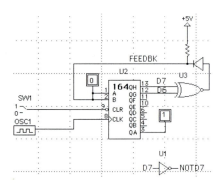

Device or signal names can be moved after they have been placed, by dragging with the pointer. They can also be edited by clicking in an existing name with the pencil cursor.

Device names are used not only as labels on the schematic, but also in component lists and bills of materials.

Setting Device Attributes

Windows—Right-click the Clock device.

Macintosh—Hold down the ⌘ key while clicking and holding on the Clock device.

◆ In the pop-up menu, select the Attributes command.

LogicWorks has a fixed number of different attribute "fields" that you can use to store auxiliary device information like component values.

◆ Click on the Value item in the field list.

◆ Type the value "14.288 MHz" in the text box, as shown.

Attributes Dialog		
Delay.Dev		**Done**
Name		
Spice		**Cancel**
Value		
		☒ Visible
Attributes for: OSC1		
14.288 MHZ		

◆ Verify that the Visible option is turned on.

◆ Click the Done button.

Notice that the value that you typed in has now appeared adjacent to the selected device. This text can be moved to any desired location relative to the device.

Pin Numbering

◆ Select the text cursor again and click it on a signal line very close to a device symbol, as shown. A blinking insertion marker will appear immediately next to the device.

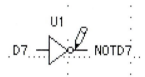

◆ Type up to four digits or letters for the pin number, then press the (ENTER) key or click the mouse button once.

Pin numbers can only be positioned directly on device pins and cannot be moved. Pin numbers are used to distinguish device pins when a netlist is created. Most non-gate library devices have pin numbers already defined for the most common package type. The default pin numbers can be specified in the device library, or they can be edited right on the diagram.

Placing Text on the Diagram

The Text cursor can also be used to place free text notations anywhere on the diagram.

◆ Click on the diagram with the text tool, not near any device or signal line, and a blinking insertion marker will appear at that point.

◆ Type any desired text, using hard returns to create multiple lines.

◆ Click the mouse button once to terminate the text item.

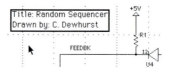

◆ Return to the arrow cursor, then click on the text item you just typed to select it.

◆ Select the Get Info command from the Schematic menu. The dialog box that appears allows you to select several framing options and set the text format for the selected item.

Creating a Bus

A "bus" is a single line on the schematic that represents a group of related signals.

◆ Using the editing techniques we have covered so far, move the XNOR, NOT, and Probe devices to the right as shown. Exact position is not important.

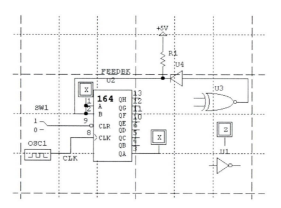

◆ Remove all lines and signal names connecting to the outputs of the 164. Also remove the label from the NOTD7 device. This will leave room to create the bus connections.

◆ Select the New Breakout command in the Schematic menu.

A "breakout" is a special symbol that indicates connections between regular signal lines and bus lines. The only way to connect a signal line into a bus is to use a breakout.

◆ Type the text "D0..7"(zero, not the letter *0*) into the pin list box as shown.

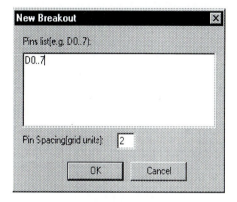

◆ If necessary, change the Pin spacing to match the illustration.

◆ Click the OK button.

The text "D0..7" tells LogicWorks to create a breakout with eight signals, D0, D1, D2, etc. up to D7. The signals don't have to be sequentially numbered like these. You can also type any collection of names separated by spaces or commas, for example: "CLK ENABLE CTRL."

◆ Place the new breakout symbol as shown so that its pins just make contact with the eight outputs of the 164. You will have to use the left arrow key on the keyboard to orient it in the direction shown.

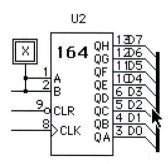

◈ Click the spacebar to reactivate the arrow cursor.

◈ Drag the breakout symbol a little to the right to make more room for the signal names and pin numbers. This is done by clicking in the diagonal line area of the breakout symbol.

A breakout can be treated just like a device symbol for editing purposes. The diagonal line area is the breakout symbol. The wide "backbone" of the breakout symbol is actually a bus line. You can now extend the bus in either direction from the breakout, just by clicking and dragging at either end of the bus line.

◈ Use the New Breakout command three more times to create small breakouts as shown. For the top one, type "D6 D7" or "D6..D7." For the middle one, type just "D0" and for the bottom one, type "D7."

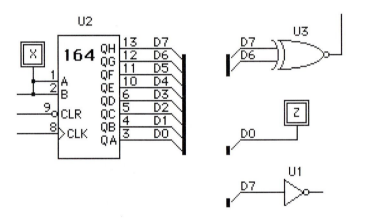

IMPORTANT: Signals are *not* connected by name between different busses. At this point we have three *different* signals called D7 that are not connected together. They will become connected together in the next step when we join the busses.

◈ Using the arrow cursor, join the bus connections of all the breakouts together, as shown. Connect one breakout to the next—do not try to connect all three at once

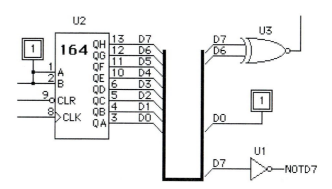

◈ Verify that the circuit still simulates correctly now that the connections are made through busses instead of directly. You may need to use the Clear X button in the Simulator Palette (or the Macintosh Clear Unknowns button) to remove X values that appeared when we broke some of the connections.

◈ Save the file you have created to this point for use later on.

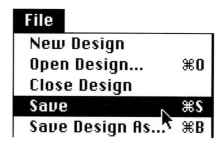

This ends the tutorial section on general schematic editing.

Digital Simulation

This section of the tutorial will provide you with a closer look at the integrated digital simulator in LogicWorks, including the following topics:

- Types of devices simulated
- Controlling the simulation
- Representation of time and signal values
- Using the trigger
- Using the signal probe.

Logic States

◆ Create a new design using the New (Windows) or New Design (Macintosh) command in the File menu.

◆ Create the partial circuit below using the O.C. Buf-1 and Binary Probe devices. (If the demolib.clf library is not already selected in the Parts Palette, select it from the drop-down list to find these devices.)

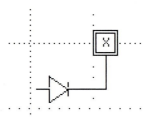

LogicWorks uses a total of 13 different logic values for signals in order to handle different drive levels and unknown situations. The probe will display an X for any of the six possible "Don't Know" states. In this case, the X results from the fact that the device input is unconnected.

◆ Add a Binary Switch device to the input of the buffer, as shown.

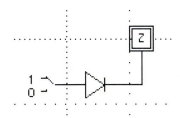

◆ Click on the switch a couple of times and note that the buffer output alternates between the 0 and Z states.

The Z value indicates a high-impedance or undriven line. Multiple open-collector or three-state devices can drive a line to simulate bus or wired-AND logic.

Circuits with Feedback

◆ Click on the buffer device and use the Duplicate command on the Edit Menu to create another one as shown.

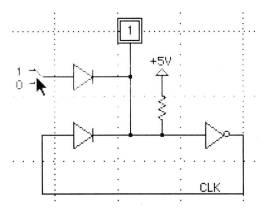

◆ Add a pullup resistor (using the Resistor and +5V symbols) and an inverter ("NOT" in the demolib.clf library), and wire them as shown.

◆ Click on the switch and notice the oscillation that occurs due to the feedback in this circuit.

◆ Name the output signal CLK so that it shows in the Timing window.

Using the Signal Probe Tool

◈ Click on the signal probe tool in the Tool Palette.

◈ Click the tip of the probe tool along any signal line. It will show the current value of the signal as the simulation progresses.

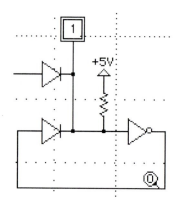

You can also use this tool to enter new signal values by typing 0 or 1 on the keyboard while the mouse button is pressed. Stuck-at, unknown, and high-impedance levels can also be inserted.

Time Values

LogicWorks uses integers to represent simulated time values. The interpretation of these numbers is left to the user, although it is usually convenient to think of them as nanoseconds.

LogicWorks uses an *event-driven* simulator, meaning that device values are only recalculated when an input change occurs. Thus, the speed at which the simulation occurs does not depend on delay or other time values in the circuit.

Primitive Devices

◆ Click on the inverter device with the arrow tool to select it. Then:

Windows—Pull down the Simulation menu and select the Simulation Params command.

Macintosh—Pull down the Simulate menu and select the Simulation Params command.

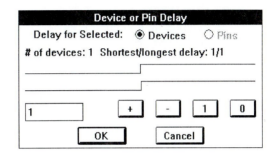

The inverter is classified as a *primitive* device since its simulation function is built into the program. Primitive devices have a single integer that defines the delay from any input pin to any output pin for any transition. More complex models can be implemented by building subcircuit devices out of the existing primitives.

◆ Click in the delay value box and change the number to 5 units, then click on the OK button. Notice the effect this delay change has on the period of the oscillation in this circuit.

Power and Ground Signals

◆ Select a 161 4-bit counter device from the demolib.clf library and place it in the circuit diagram as shown.

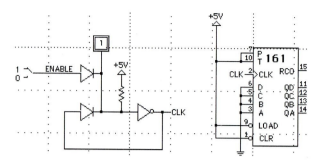

◆ Using the text cursor, as described previously, add the names "CLK" and "ENABLE."

◆ Place +5V and Ground symbols as shown to permanently fix these signals to high and low levels, respectively.

Subcircuit Devices

◆ Add the names D0 to D3 using the following procedure:

   ■ Name the least significant counter output D0 using the usual technique.

   ■ Hold down the (CTRL) key on the Windows keyboard (or the ⌘ and (OPTION) keys on the Macintosh keyboard) while you click on each higher pin in turn. Make sure you click only at the very end of the pin. This will automatically place sequential numbers on the lines clicked.

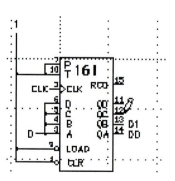

Notice that the traces D0 to D3 in the Timing window will show unknown values because the counter has never been cleared into a known state.

Windows—Press the Clear Unknowns button in the Simulator Palette.

Macintosh—Press the Clear X button in the Simulator Palette.

This resets all storage elements to the zero state and clears unknown lines.

◆ Reactivate the arrow cursor.

Windows—Click on the 161 device to select it, then select the Parameters command in the Options menu of the Simulator Palette.

Macintosh—Click on the 161 device to select it, then select the Simulation Params command in the Simulation menu.

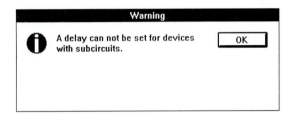

The 161 counter is a *subcircuit device*, meaning that its logic function is implemented using a combination of the LogicWorks primitive devices. Because of this, the overall delay for the device cannot be adjusted by simply changing one parameter. Two methods are available for modifying delays in subcircuit devices and these are discussed in the following sections.

◆ Click the OK button on the warning box.

Windows—

◆ Right-click on the 161 device.

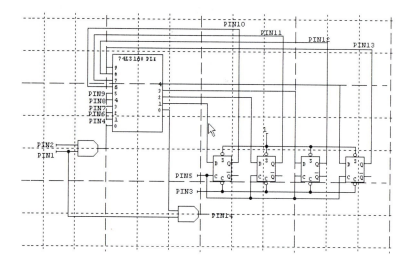

◆ In the pop-up menu, select the Device Info command.

◆ Click on the "Lock Opening Subcircuit" check box, to turn it off.

◆ Click on the OK button to close the dialog.

◆ Double-click on the 161 device to open its internal circuit.

A new window will open showing the internal circuit of this device. Notice how you can use the signal probe tool, the Parameters command, and all the drawing tools to view and modify this internal circuit. If you modify this circuit, *all devices of the same type* in this design will be equally affected.

◆ Double-click on the Control Menu icon at the upper left corner of the window to close the internal circuit.

Macintosh—

◆ (COMMAND)–click on the 161 device.

◆ In the pop-up menu, select the Device Info command.

◆ Click on the "Lock Opening Subcircuit" check box, to turn it off.

◆ Click on the OK button to close the dialog.

◆ Double-click on the 161 device to open its internal circuit.

A new window will open showing the internal circuit of this device. Notice how you can use the signal probe tool, the Simulation Params command and all the drawing tools to view and modify this internal circuit. If you modify this circuit, *all devices of the same type* in this design will be equally affected.

◆ Click the close box at the upper left corner of the window to close the internal circuit.

Pin Delays

◆ Using the arrow cursor, click midway along the QA output pin on the 161 device to select it.

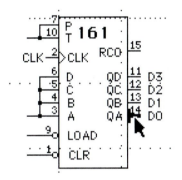

Windows—Select the Parameters command from the Simulator Palette's Options menu.

Macintosh—Select the Simulation Params command from the Simulate menu.

LogicWorks allows you to set a delay on an individual pin on a primitive or subcircuit device. The logical effect is the same as if you had inserted a buffer device with the specified delay in series with the pin. Pins always have a default delay of zero.

◆ Set the pin delay to 3 units and click OK. Notice the effect this has on the D0 trace in the Timing window.

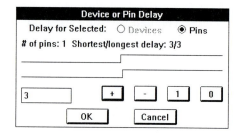

Pin delays can be used to customize path delays in subcircuit devices without opening and modifying their internal delays. Setting pin delays on a subcircuit device affects only the single device modified, whereas changing internal primitive device delays will affect all copies of the same type of device.

Moving Timing Traces

◈ Click and vertically drag the name CLK in the Timing window and reposition it relative to the other traces.

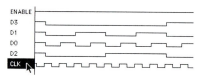

You can reposition any group of traces for ease in making timing comparisons. Any number of traces can be moved at once by holding the (SHIFT) key while clicking on the trace names.

Grouping Timing Traces

◈ Click on the name D0 in the Timing window. Hold the (SHIFT) key depressed while you click on names D1, D2, and D3 so that they are all selected.

Windows—Press the right mouse button on any of the four selected names.

Macintosh—Hold down the ⌘ key while you click and hold on any of the four selected names.

◆ In the pop-up menu, select the Group command.

You will now see that the four traces D0 to D3 collapse into a single grouped trace showing their combined value in hexadecimal. The same pop-up menu can be used to Ungroup the signals again, or to set the signal order used to create the hexadecimal value.

Note that the grouped trace has double vertical bars on some values. This is due to the delay we inserted in the QA output pin. If you set the pin delay back to zero, this will disappear.

Using the Trigger

Windows—Click on the Trigger button (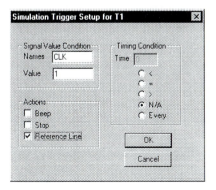) in the Simulator toolbar.

Macintosh—Click on the Trigger button on the Timing control panel. (If the control panel is not showing, select the Show Timing Palette command from the Simulate menu).

The trigger mechanism allows you to detect various timing and signal-state conditions.

Windows—Type the name CLK in the Names box.

Macintosh—Type the name CLK in the Signals box.

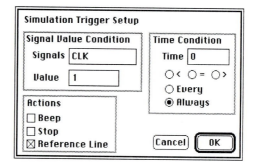

◆ Type the value 1 in the Value box.

◆ Select the Reference Line option.

◆ Click the OK button.

You will now see that a reference line is drawn on the Timing window each time the CLK signal changes to a 1 state. (If these lines do not show up, make sure the simulator is running.) You can also enter ranges of signal names (e.g. D7..0) and corresponding hexadecimal values (e.g. 7A) into these boxes to match more complex events.

Creating Device Symbols

This tutorial will show you how to use the DevEditor tool to create your own device symbols.

Creating a New Library

Windows—

◆ Using the right mouse button, click on the Parts Palette. Select the New Lib command in the pop-up menu.

◆ Create a new library called mylib.clf in the LogicWorks directory.

Macintosh—

◆ ⌘–click on the Parts Palette and select the New Lib command in the
   pop-up menu.

◆ Create a new library called mylib.clf in the LogicWorks folder.

Device library files hold collections of part symbols along with associated
pin function information, default attribute values, and internal circuit definitions. A single library can contain from one to thousands of part definitions, to suit your needs.

Creating a Device Symbol

Windows—Click on the New Document button ( ) in the toolbar, then select
   Device Symbol and click OK A new device symbol editor window and the
   associated toolbars will appear:

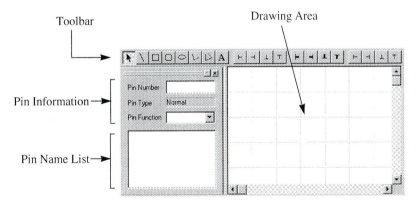

Macintosh—Select the DevEditor command from the Tools menu.

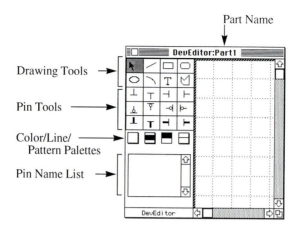

The DevEditor window contains a drawing area for your symbol, plus a
Tool Palette and a Pin Name List. The Tool Palette includes standard drawing tools plus special items for normal, inverted, and bus pin placement.

◆ Click on the polygon (◁) tool in the tool palette.

◆ Draw a symbol similar to the one shown by clicking once at each
corner point and then double-clicking to terminate the polygon.

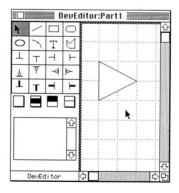

The position of the symbol in this window is not important.

◆ Select the (⊣) pin tool.

◆ Place input pins on the symbol by clicking at the positions shown.

◆ Select the (⊢) pin tool.

◆ Place an output pin as shown.

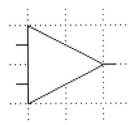

NOTE: The crossbar portion of the T pin tool only appears during placement and dragging, for alignment purposes.

Entering Pin Names and Numbers

Placing the pin graphics on the symbol will have automatically added pin names to the pins list. In this section we will enter more meaningful names and add pin numbers.

Windows—

◆ Double-click on the PIN1 item in the pin list to open the name for editing.

◆ Type the new name "INA" and type Enter.

◆ Click in the Pin Number box and type "2."

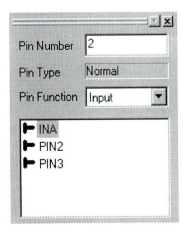

◆ Similarly, select the PIN2 item and enter pin number "3."

◆ Select the PIN3 item and enter pin number "6."

Macintosh—

◆ Double-click on the PIN1 item in the Pin List (on the left edge of the screen).

◆ In the Pin Information Palette that appears, enter the pin name "INA" and pin number "2."

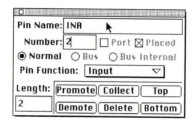

◆ Press the (ENTER) key on the keyboard to move to the next pin in the list.

◆ Enter name "INB" and pin number "3."

◆ Press (ENTER) again to see the last pin and enter pin name "OUT" and number "6."

◆ Press (ENTER) to complete the last pin, then close the Pin Information box.

We have now entered default values for the pin numbers that will appear in a netlist. These can be edited on the schematic for individual pins, if desired.

Saving and Using the Part

Windows—Click the Save button { 🖫 } on the toolbar.

Macintosh—Select the Save Part As command in the File menu.

◆ Enter the part name "LM741" or any other desired name.

◆ Double-click on the mylib.clf library in the list to select the destination.

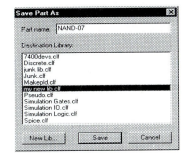

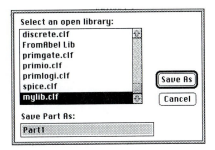

◈ Close the DevEditor window.

◈ If the mylib.clf library is not already selected, select it in the library drop-down list in the Parts Palette.

◈ Double-click on the newly-created part and place a copy of it in the schematic.

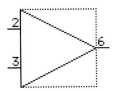

Auto-Creating a Symbol

For standard types of rectangular symbols, the Autocreate feature will generate a symbol for you in seconds.

Windows—Click on the New Document button ( ) in the toolbar, then double-click Device Symbol.

Macintosh—Select the DevEditor command from the Tools menu.

◆ Select the Auto Create Symbol command.

◆ In the "Left Pins" box, type the text "D7..0(9..2),,,CLK(1)."

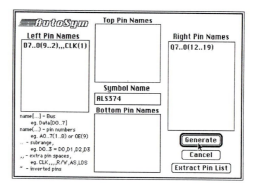

The entry "D7..0" will generate a set of eight pins named D7, D6, etc. "(9..2)" are the corresponding pin numbers. The three commas indicate that we want extra space between these pins. "CLK(1)" creates a single pin called CLK with pin number 1. The pin numbers can be omitted, if desired.

◆ In the Right Pins box, type the text "Q7..0(12..19)."

◆ In the Name box in the center, type "ALS374 " or any other desired symbol name.

◆ Click the Generate button.

The auto-generated symbol should now display the pins and pin numbers entered above. These items can be edited using the drawing tools and Pin Info box, if desired.

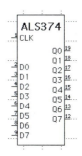

◆ Save the new part to the mylib.clf library, as described earlier.

◆ Close the DevEditor window.

The symbol can now be placed in a schematic in the usual manner.

5
Schematic Editing

This chapter describes the elements of a LogicWorks circuit design and the procedures you can use to create one.

Design Structure

What is a Design?

In LogicWorks, the term "design" refers to a complete, independent set of circuitry, including all the information needed to display, edit, and simulate it. The following rules outline how a design is stored:

- A single design is stored in a single file and no logical connections are made between designs. All information required to display and edit a design is stored in the design file.

- A design never makes reference to external library files. When a symbol is used from a library, all information needed is read from the library and stored with the design. Changing the original library definition *will not* automatically update the design.

- A design has a top-level circuit, referred to as the *master* circuit. This circuit may contain any number of devices which themselves can contain circuits, called *subcircuits*. Subcircuits can be nested to any desired depth, limited only by available memory. Many LogicWorks operations, such as text report generation, apply only to the master circuit.

- When a design file is opened, the entire contents of the design are read into memory. This means that design sizes are limited by the available memory in your computer and increasing the memory allocated to the program will increase the size of the designs you can work with.

■ A number of user-selectable parameters are stored with the design and affect the entire design when changed. These include:

   ■ Attribute and pin number text style settings;

   ■ Display options, such as crosshairs and printed page breaks;

   ■ Printer page setup.

What is a Circuit?

In LogicWorks, the term "circuit" refers to a single circuit page, as displayed in a single window.

■ Each master circuit or subcircuit consists of one page.

■ Each circuit is viewed in a separate circuit window, and any number of circuits or subcircuits can be displayed on the screen simultaneously.

■ A circuit page is drawn on the screen as if it were a single piece of paper, although it may have to be broken up into a number of individual sheets of paper for printing or plotting.

■ If the circuit is a subcircuit, then logical connections to the parent device symbol are made using the Port Connector device. Port connectors in the subcircuit are matched by name with pins on the parent device.

◆ See more information on subcircuits and port connectors in Chapter 6, Advanced Schematic Editing.

Types of Objects in a Circuit

A LogicWorks circuit is made up of three types of entities: devices, signals, and text objects.

■ A device is an object having a symbol, signal connection points called "pins," and optional attributes, internal circuit, and simulation information. A device in LogicWorks can correspond to a physical device in a circuit, or it can be a *pseudo-device*, such as a Ground connector or bus breakout which is used for schematic notation purposes.

■ A pin is a connection point on a device. A pin is not an independent entity, since it only exists as part of a device and cannot be created or deleted separately. However, pins can have attributes, pin numbers, and other parameters that may be different from pin to pin on the same device. The Get Info command can be used on a selected pin to view and set pin parameters. A *bus pin* is a special type of pin that represents an arbitrary number of internal pins. The internal pins are not visible on the schematic but can still have the same logical properties as other pins.

■ A signal is a conductive path between device pins. Signal connections can be made visually by drawing lines between device pins, or logically by name or bus connection.

■ A text object is used to place a title block or other notation on the diagram. Text can be typed and edited directly within LogicWorks, or can be created externally and pasted onto the diagram from the Clipboard. Text objects are not associated with any other object and are not accessible through net or component lists. The attribute facilities should be used to associate text with specific devices or signals.

■ A picture object is used to place a border, logo, or mechanical drawing on the diagram. Picture objects can be created externally and pasted onto the diagram from the Clipboard or created using the device symbol editor. Once placed, a picture object becomes a single element that can be moved, duplicated, and deleted, but it cannot be edited within LogicWorks.

Design Operations

This section describes how to work with LogicWorks circuit designs.

Creating a New Design

Windows—To create a new design, click on the New Document button (□) in the toolbar, or select the New item in the File menu, then select Design from the list of document types.

Macintosh—To create a new design, select the New Design command in the file menu.

The new design will consist of an empty master circuit that will appear in a Windows circuit window as Design1.CCT, Design2.CCT, and so on, and in a Macintosh circuit window as Design1, Design2, and so on. This command does not create a disk file. The design exists only in memory until you save it using the Save As command.

Your circuit diagram is created by first placing one or more devices in the circuit window (as described below), and then interconnecting the device pins with signal connections.

Closing a Design

A design is closed when its master circuit window is closed.

Windows—At the top left corner of the master circuit's window, click on the X icon or select the Close command from the File menu.

Macintosh—Click once on the close box or select the Close Design command in the File menu.

In either case, you will be prompted to save the design before closing if any changes have been made.

Disposing of a Design File

LogicWorks has no built-in command to dispose of a design file. All information about a design is stored in a single file. You may, however, simply delete this file via your computer's operating system (the Windows Explorer or the Macintosh Finder).

Navigating Around a Schematic

In addition to the standard scroll bars and Reduce/Enlarge menu items, LogicWorks has a number of convenient features for moving around a diagram.

Auto-Scrolling

Whenever the mouse button is depressed and moves close to the edge of a Schematic window, the window automatically scrolls to expose more area on that side.

Zoom (Magnifying Glass) Tool

The ⌕ item in the Tool Palette is a powerful tool for moving around in a schematic diagram. Once you have activated this tool, it can be used to zoom in and out, and to control the exact area displayed on the screen.

- Clicking and releasing the mouse button on a point on the diagram will zoom in to that point by one magnification step.

- Clicking and dragging the mouse down and to the right zooms in on the selected area. The point at which you press the mouse button will become the top left corner of the new viewing area. The point at which you release the button will become approximately the lower right corner of the displayed area. The circuit position and scaling will be adjusted to display the indicated area.

- Clicking and dragging the mouse upward and to the left zooms out to view more of the schematic in the window. The degree of change in the scale factor is determined by how far the mouse is moved. Moving a small distance zooms out by one step (equivalent to using the Reduce command). Moving most of the way across the window is equivalent to choosing the Reduce to Fit command.

The Clipboard

The standard Clipboard commands, Cut, Copy, and Paste, can be used to move or copy circuit fragments, graphical, and text information within a single circuit window, between windows, or between LogicWorks and other programs (e.g., word-processing or graphics packages).

Using Clipboard Data From Other Programs

When you start up LogicWorks, the Clipboard may contain text or graphical information cut or copied from a document in another program. Logic-Works allows you to make use of this information as follows:

■ Text information from a word processor or text editor can be pasted into a text block.

■ Picture information from other applications can be pasted onto a LogicWorks circuit diagram.

◆ See more information in the Edit menu section of Chapter 12, Menu Reference.

Using Clipboard Data From LogicWorks

When a Cut or Copy is performed, two types of data are placed on the Clipboard:

■ A bitmap picture (Windows BMP format or Macintosh PICT format) of the selected items. This could be pasted into a graphics document using most drawing programs.

■ The LogicWorks circuit info for the selected items. This data is in a format that only LogicWorks can understand, and is discarded when you Quit. This means that you cannot transfer circuit information between LogicWorks sessions.

The Cut and Copy commands work on the currently selected group of objects and will be disabled if nothing is selected. See the section below on "Selecting Circuit Objects." When items are copied onto the Clipboard, their names are copied with them, which may result in duplicate names. If duplicate signal names are pasted back into the circuit from which they were copied, then logical connections will be made between the like-named segments.

◆ See more information in the Edit menu section of Chapter 12, Menu
Reference.

Selecting Circuit Objects

Many LogicWorks commands, such as Get Info, Cut, Copy, etc., operate on
the currently selected objects. To select circuit objects, the cursor must be
in the normal Point (↖) mode.

Selecting a Device

A single device is selected by clicking the mouse button with the pointer
positioned anywhere inside the device symbol, or in any displayed attribute
value associated with the symbol.

Simulated input devices, such as switches and keyboards, can only be
selected by holding the (SHIFT) key while clicking. This is necessary because
a normal click is used to change the state of these devices.

Selecting a Text Object

A single text item is selected by clicking the mouse button with the pointer
positioned anywhere inside the item.

Selecting a Picture Object

A single picture item is selected by clicking the mouse button with the
pointer positioned anywhere inside the item.

Selecting a Signal

A single signal is selected by clicking anywhere along the signal line. This
selects only the part of the signal directly attached to the clicked line. Double-clicking the signal selects all parts of the signal, including logical connections by name or bus.

Selecting a Pin

A pin is selected by clicking on the pin line close to the device.

NOTE: Since an unconnected device pin is both a pin and a signal, you determine whether you get the pin or signal pop-up menu as follows:

Windows—Right-clicking on the pin in the last 1/4 of the pin length away from the device will display the signal menu.

Macintosh—⌘–clicking on the pin in the last 1/4 of the pin length away from the device will display the signal menu.

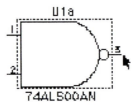

Selecting the Signal

▓   In either version, clicking on the pin close to the device symbol will display the pin menu.

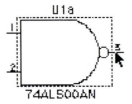

Selecting the Pin

Selecting Groups of Objects

Several methods are available for selecting multiple objects:

▓ Any group of adjacent items can be selected by activating the Point tool and clicking and dragging across the group. A flickering rectangle will follow the mouse movement. Any object that intersects this rectangle when the button is released will be selected.

▓ A group of interconnected devices and signals is selected by double-clicking on any device in the group while holding down the Windows (CTRL) key or the Macintosh (OPTION) key. If a circuit is completely interconnected, this will select the entire circuit.

▓ The Select All command in the Edit menu selects all items in the current circuit design.

▓ The (SHIFT) key can be used in combination with any of the above methods to select multiple items. When the (SHIFT) key is held, the previously selected items remain selected when a new item is clicked on. Thus you can add to the selected group until the desired collection of items is selected.

Changing Search Order

Holding down the Windows (CTRL) key or the Macintosh (OPTION) key while clicking the pointer causes object types to be searched in the opposite order from normal. This can be used, for example, to select a signal name that has accidentally moved under a device.

Deselecting a Selected Object

All currently selected objects are deselected by clicking in an empty area of the Schematic window. A single item can be deselected by holding the (SHIFT) key while clicking on it.

Classes of Devices

For the purposes of this section, devices in LogicWorks can be divided into four groups:

■ **Symbol-only devices:** These are symbols which are used to represent physical devices on a schematic, but which have no simulation function. For example, the analog components provided in the discrete.clf library fall into this category.

■ **Subcircuit devices:** These are symbols which have a simulation function defined by an internal circuit. The 7400 devices provided with LogicWorks fall into this category. The internal circuit for this kind of symbol can be viewed by double-clicking on the symbol.

■ **Pseudo-devices:** These are the symbols used for bus breakouts, power and ground symbols, and so on. They do not represent an actual physical device in a circuit, but they have specific meanings on the schematic diagram.

■ **Simulation primitives:** These are device symbols which have a built-in simulation function when used with the LogicWorks simulator.

◆ See a description of LogicWorks primitive types in Chapter 9, Primitive Devices.

Device Libraries

The symbols and related parameters for LogicWorks devices are stored in data files called device libraries. Libraries can be opened and closed by displaying the Parts Palette's pop-up menu and using the Open and Close commands, or by using entries in the Windows initialization file or the Macintosh setup file.

For each device symbol in a library, the following data is stored:

■ General information on the type, such as number of pins, number of inputs, number of outputs, type name, default delay, default attributes, position, orientation and type of each pin, and so on.

■ A picture representing the symbol for this type.

■ An optional internal circuit definition.

How Device Symbols are Created and Stored

Libraries are created and modified using the DevEditor tool, which is described elsewhere in this manual.

◆ See Chapter 11, Device Symbol Editing, for more information.

Placing and Editing Devices

Selecting a Device From a Library

To select a device from a library for placement in the schematic:

◆ Macintosh Only—Select the desired library using the drop-down list at the top of the Parts Palette.

◆ If necessary, use the scroll bar to scroll the library's parts list until the desired part name is in view.

◆ Double-click on the part name in the list.

◆ Move the cursor to the current Schematic window.

The cursor will be replaced by an image of the selected device. While moving this flickering image around, you can use the arrow keys on the keyboard, or the orientation tools on the Tool Palette, to rotate the symbol.

Clicking anywhere in the circuit diagram will make a permanent copy of the flickering device at that point.

NOTE: Holding down the Windows (CTRL) key or the Macintosh (OPTION) key while clicking will inhibit checking for pin connections. This allows you to select the device again and drag it to a new position without affecting any existing connections.

Duplicating an Existing Device

To duplicate an existing device on the schematic, either:

■ Select a similar device anywhere on the current circuit and use the Duplicate command (either in the Edit menu or in the device pop-up menu); or

■ Select a similar device in any other open circuit window and use the Copy command. Return to the destination circuit window and select the Paste command.

After either of these operations, the cursor will be replaced by a flickering image of the selected device. This copy can be placed by clicking in the schematic, as discussed earlier.

Deleting a Device

Devices can be removed by either of two methods:

■ Select the device by clicking on it (holding the (SHIFT) key if it is a switch or other input device). Then press the (BACKSPACE) or (DELETE) key on the keyboard, or select the Clear command from the Edit menu. Or:

■ Enter Zap mode, by selecting the Zap command on the Edit menu or clicking on the Zap icon in the Tool Palette. Then click on the device in question.

Moving a Device

Devices can be moved by clicking and dragging them to the desired new position. If more than one device is selected, all the devices, and all signals connecting between them (whether or not selected), will be moved. Signal lines will be adjusted to maintain right angles at points where moving signal lines intersect with non-moving ones.

Entering Device Attributes

To enter device attributes, either:

■ Display the device's pop-up menu (Windows users right-click on the device; Macintosh users ⌘–click on the device). Then select the Attributes command, or do the following:

■ Select the device by clicking on it normally. Then choose the Get Info command from the Schematic menu, and click the Attributes button.

◆ See more information on entering and using attributes in Chapter 6, Advanced Schematic Editing.

Drawing Signals

Signal lines are drawn in either Point (➤) mode or Signal Drawing (+) mode.

Interconnecting Signals

If you draw a signal line so that the end of the line makes contact with a second signal line, then those two signals will be interconnected. Also, if you place a new device so that one of its pins touches an existing signal line, that pin will be connected to the signal. If both of the signals being connected were named, then you will be prompted to choose the name of the resulting signal. Whenever three or more line segments belonging to the same signal meet at a given point, an intersection dot will be placed at that point automatically.

NOTE: For efficiency, signals are only checked for connections at their endpoints and only signals actively being edited are checked. It is possible to create overlapping lines that do not connect by unusual combinations of editing operations. This situation is usually visually apparent at the time the editing is done, since the intersection dot will be missing and the entire signal will not highlight when clicked on.

◆ See more information on connection-checking under the Paste command in Chapter 12, Menu Reference.

Connecting Signals by Name

◆ See the section below on Name and Pin Number Operations for details on how signals are connected by name.

Signal Line Editing

Drawing from an Existing Line or a Device Pin

A line can be extended from the end of an existing line or device pin using the arrow (↖) cursor. Click and hold on the end of the pin and drag away from the pin. A pair of right-angle lines will follow the cursor away from the pin as long as the mouse button is pressed. Releasing the mouse button makes these lines permanent. If the end of the line (i.e., the point where the mouse button was released) touches another signal line, a connection will be made at that point.

Windows—Alternate line-routing methods can be activated by pressing the (CTRL) and (TAB) keys, as follows:

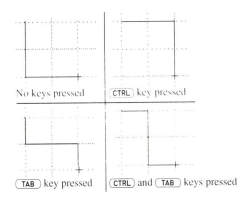

| No keys pressed | (CTRL) key pressed |
| (TAB) key pressed | (CTRL) and (TAB) keys pressed |

The (CTRL) key inverts the order of line drawing, and the (TAB) key switches to three line segments with a center break. The (SHIFT) key constrains the movement to a single vertical or horizontal line.

NOTE: Holding the (CTRL) key while clicking will inhibit checking for pin connections. This allows you to select the signal again and drag it to a new position without affecting any existing connections.

Macintosh—Alternate line-routing methods can be activated by pressing the (OPTION) and ⌘ keys, as follows:

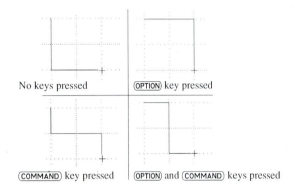

The (OPTION) key inverts the order of line drawing, and the (COMMAND) key switches to three line segments with a center break. The (SHIFT) key constrains the movement to a single vertical or horizontal line.

NOTE: Holding down the (OPTION) key while clicking will inhibit checking for pin connections. This allows you to select the signal again and drag it to a new position without affecting any existing connections.

Creating an Unconnected Signal Line

The Draw Sig (+) tool can be used to create an unattached signal line, or to extend an existing signal. Simply click anywhere in the schematic and drag in the desired direction. Unlike the Point mode drawing method, above, the mouse button does not have to be held down while creating signals in this mode. Double-clicking terminates the signal line.

Editing a Signal Line

The following features are available to edit signal lines:

■ Zap mode (entered by selecting the Zap command in the Edit menu or the Zap item in the Tool Palette) allows you to remove any single line segment from a signal connection. Zapping on a signal line removes only the line segment to which you are pointing—up to the nearest intersection, device pin, or segment join point.

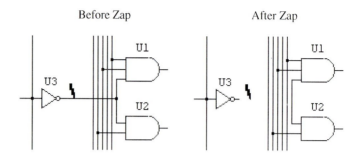

Before Zap   After Zap

- Selecting a signal line (by clicking anywhere along its length), then hitting the (DELETE) key or selecting the Clear command from the Edit menu, removes an entire signal trace.

- Drawing backwards along the length of an existing line causes the line to be shortened to end at the point where you let the button go.

- Clicking and dragging the middle of a signal line segment allows you to reposition the line. Vertical lines can be moved horizontally and vice versa.

Checking Signal Interconnection

Double-clicking anywhere along a signal line will cause that signal segment and all logically connected segments to be selected.

Name and Pin Number Operations

Names may contain any letters, numbers, or special characters that you can type on the keyboard, but are restricted in length to 15 characters. The name associated with an object can be placed anywhere on the diagram, and will be automatically removed if the object is removed.

Pin numbers may contain at most 4 characters.

Naming Signals and Busses

What Signal Names are Used For

The signal name is referenced by the following LogicWorks functions:

▩ The signal name is used in Report Generator output, such as netlists.

▩ Signals can be logically interconnected by name.

▩ Signal names are used to identify traces in the Timing window.

Adding a Signal Name

Windows—To name a signal, enter Text mode, either by selecting the Text command in the Edit menu, or by clicking on the text icon in the toolbar:

Text Tool

Macintosh—To name a signal, enter Text mode, either by selecting the Text command in the Edit menu, or by clicking on the text icon in the Tool Palette:

Text Tool

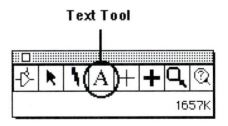

Note that once "Text" is selected, the cursor changes to a pencil icon.

Press and hold the mouse button with the tip of the pencil positioned anywhere along a signal line except within five screen pixels of the device. As long as you hold down the mouse button, an I-beam cursor will track the mouse movements. The signal-name text will start at the position where you release the button. Type the desired name, and press (ENTER) or click the mouse button anywhere.

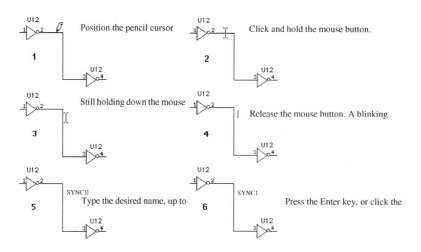

Multiple Naming of Signals

A signal name can appear in up to 100 positions along the length of the same signal line. To add a new position, simply use the normal naming procedure given in the section on signal naming, such as:

◆ Select Text mode.

◆ Click and drag anywhere along the signal line.

◆ Release the mouse button.

A new copy of the signal's name will appear at this point, followed by a flashing cursor. To accept the name, simply click the mouse button once or press the (ENTER) key. If you edit any occurrence of a name along a signal segment, all other occurrences will be updated to reflect the new name.

Any occurrence of a signal name can be removed using the Zap tool. If you remove the last visible name from a signal segment, then the logical connectivity to other like-named signals is removed.

Connecting Signals by Name

Signal names can be used to make logical connections between lines that are not visually connected on the schematic. The following rules apply:

▪ Signal names must be visible to be checked for connections, unless a Signal Connector device (such as Ground) is attached. More information on invisible names is given in the following section.

■ Signal names are known throughout a schematic page. Like-named signal lines are thus logically connected for simulation and netlisting purposes. Whenever a signal name is added or changed, the circuit is checked for a change in connectivity. If the name is now the same as another signal, the two signals are merged into one. If this signal segment was previously connected by name to others, and the name is changed, then the logical connection is broken. Whenever a name change causes two signals to be connected, the changed signal will flash on the screen to confirm the connection.

■ Signals which are contained in busses are a special case. Every signal contained in a bus has a name, even if it is not displayed on the diagram. However, the names of bussed signals *will not* be used to make logical connections unless an explicit name label has been added to the signal line.

For example, if you have a bus containing a signal named CLK and a separate signal line also named CLK, there will be no logical connection between these two signals. The name appearing on the bus breakout is part of the breakout symbol and is not considered to be a name label. If an explicit label is added to the bussed CLK signal (using the text cursor) then the two CLKs will be logically connected.

■ The same rules discussed above for signals also apply to busses. Whenever two busses are logically connected, all like-named internal signals also become logically connected.

Device Names

In this book, we use the term "device name" to refer to the character string that identifies a unique device in the circuit. Typical device names might be U23, C4, IC12A, and so on. This is distinct from the type name or part name that is used to distinguish the type definition that is read from a device library. Typical part names are 74LS138, MC68000L8, SPDT Switch, and so on.

Adding a Device Name

Enter Text mode either by selecting the Text menu item in the Edit menu, or by clicking on the text icon in the Tool Palette:

Text Tool

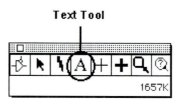

Once Text mode is selected, the cursor changes to a pencil icon. Press and hold the mouse button with the tip of the pencil positioned inside a device symbol. As long as you hold down the mouse button an I-beam cursor will track the mouse movements. The device-name text will start at the position where you release the button. Type the desired name, and press (ENTER) or click the mouse button anywhere.

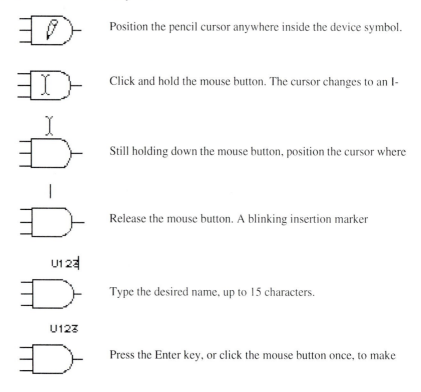

Position the pencil cursor anywhere inside the device symbol.

Click and hold the mouse button. The cursor changes to an I-

Still holding down the mouse button, position the cursor where

Release the mouse button. A blinking insertion marker

Type the desired name, up to 15 characters.

Press the Enter key, or click the mouse button once, to make

Once a name is placed, it can be repositioned by dragging it using the arrow cursor, or removed using the Zap cursor. The device name will be removed automatically if the device is removed.

Adding an Invisible Name

An invisible name for either a device or signal can be created in one of two ways.

Windows—

■ Use the right mouse button to select the device or signal, then select the Name command from the pop-up menu. Or:

■ Select the desired device or signal, then select the Get Info command in the Schematic menu (CTRL–I), then click on the Attributes button in this dialog, then select the Name field in the Attributes Dialog.

In either case, if the name is already visible on the diagram, changing it here will change all its displayed occurrences.

Macintosh—

■ Hold the ⌘ key while clicking on the device or signal, then select the Name command from the pop-up menu. Or:

■ Select the desired device or signal, then select the Get Info command in the Schematic menu (⌘–I), then click on the Attributes button in this dialog, then select the Name field in the Attributes Dialog.

In either case, if the name is already visible on the diagram, changing it here will change all its displayed occurrences.

IMPORTANT: When a signal name is invisible, it is *not* used to establish connections by name to other signal lines. See the rules in the section above, Naming Signals and Busses.

Making an Invisible Name Visible

An invisible name can be made visible by either of the following methods:

■ Click the Text pointer anywhere on the signal or device. When the mouse button is released, the name will be positioned at that point, as described in the general naming instructions above. Or:

■ Select the Name command in the device or signal pop-up menu, and enable the Visible option. The name will be displayed in a convenient location close to the object.

Auto-Naming Features

Windows—Three features are available to simplify the naming of groups of related signals, devices, and pins. These features are activated by holding down the (CTRL), (SHIFT), and/or (TAB) keys, then selecting the signal to be named with the Text cursor.

■ Auto-alignment—If the (TAB) key is held down while the signal is selected, the text insertion point will be positioned horizontally aligned with the last signal name that was entered. The vertical position is determined by the vertical position of the line that was clicked on. This feature works only with signal or device names, not with pin numbers.

■ Auto name generation—If the (CTRL) key is held down while a signal, device, or pin is selected, a new name is generated automatically for this item. The new name will be the same as the last one entered, except that the numeric part of the name will have been incremented. If the previously-entered name did not have a numeric part, then a "1" digit will be appended to it. If the (SHIFT) key is pressed at the same time, the number will be decremented instead of incremented.

Sequential Naming

The above two features can be used in combination to perform easy naming of sequential signals. The normal symbol standard in LogicWorks is to position the highest numbers at the top, so you can either:

■ Number the topmost line in the group (e.g., D7) using the normal naming technique, described above. Then hold down the (TAB), (CTRL), and (SHIFT) keys while clicking on successive lower-numbered lines. Or:

■ Number the bottom-most line in the group (e.g., D0) using the normal naming technique, described above. Then hold down the (TAB) and (CTRL) keys while clicking on successive higher-numbered lines.

Note that when you select each successive line, the new name appears; however, it is necessary to click again (or press (ENTER)) to make the name permanent.

Macintosh—Two features are available to simplify the naming of groups of related signals, devices, and pins. These features are activated by holding down the ⌘, (OPTION), and/or (SHIFT) keys, then selecting the signal to be named with the Text cursor.

- **Auto-alignment:** If the ⌘ key is held down while the signal is selected, the text insertion point will be positioned horizontally aligned with the last signal name that was entered. The vertical position is determined by the vertical position of the line that was clicked on. This feature works only with signal or device names, not with pin numbers.

- **Auto name generation:** If the (OPTION) key is held down while a signal, device, or pin is selected, a new name is generated automatically for this item. The new name will be the same as the last one entered, except that the numeric part of the name will have been incremented. If the previously-entered name did not have a numeric part, then a "1" digit will be appended to it. If the (SHIFT) key is pressed at the same time, the number will be decremented instead of incremented.

Sequential Naming

The above two features can be used in combination to perform easy naming of sequential signals. The normal symbol standard in LogicWorks is to position the highest numbers at the top, so you can either:

- Number the topmost line in the group (e.g., D7) using the normal naming technique, described above. Then hold down the ⌘, (OPTION), and (SHIFT) keys while clicking on successive lower-numbered lines. Or:

- Number the bottom-most line in the group (e.g., D0) using the normal naming technique, described above. Then hold down the ⌘ and (OPTION) keys while clicking on successive higher-numbered lines.

Note that when you select each successive line, the new name appears; however, it is necessary to click again (or press (ENTER)) to make the name permanent.

Removing a Name

A device or signal name can be removed by using the Zap pointer, as described in the section on Deleting a Device, above. If the signal has been named in multiple locations, then Zap removes the name only at the location zapped.

Editing a Name

The name can be changed by simply clicking the Text pointer on the signal name and editing it using the keyboard. Alternatively, a name can be edited by choosing the Name command in the pop-up menu for the device or signal. Changing the name in the resulting dialog—or at any single location on the diagram—will change all visible occurrences of it.

Moving a Name

A device or signal name can be moved by activating the arrow cursor, clicking and holding the mouse button on the name, and dragging it to the desired new position. Pin numbers cannot be repositioned.

Setting and Editing Pin Numbers

Pin numbers may contain one to four characters. They are always positioned adjacent to the associated pin. Any characters may be used—not just digits—in order to accommodate alphanumeric pin numbering for pin grid arrays.

Uses of Pin Numbers

Pin numbers are used only for labeling purposes and have no particular connectivity significance to LogicWorks. Pin numbers are not checked for duplicates or other invalid usage. Pin numbers placed on a diagram will be used in creating a netlist (see Chapter 13, Creating Text Reports), and will appear when the circuit is printed. If a pin is unnumbered, it will appear in a netlist with a "?"—unless the device has three or fewer pins (e.g., discrete components), in which case it will be sequentially numbered.

Default Pin Numbers

A device symbol may have default pin numbers which will appear when the device is first placed. These pin numbers are not permanent and can be edited or removed by techniques discussed in this section. These default pin numbers are assigned using the DevEditor tool.

◈ See Chapter 10, Device Symbol Editing.

Editing Pin Numbers On the Schematic

In Text mode, if the mouse button is pressed with the tip of the pencil pointer positioned on a signal line within five pixels of a device, a blinking insertion bar will appear immediately where the signal joins the device. You cannot set the text position for pin numbers. Type the desired one- to four-character number, then press (ENTER) or click the mouse button to make the number permanent.

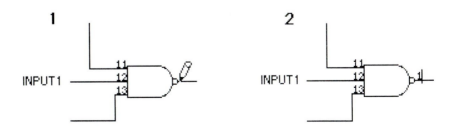

Editing Pin Numbers Using Get Info

To edit pin numbers using the Get Info dialog box:

◈ Display the device's pop-up menu: In Windows, right-click on the device. On the Macintosh, ⌘–click on the device.

◈ From the pop-up menu, choose Device Info.

◈ In the Device Info dialog, click on the Pin Info button. This will display the pin information for the first pin.

◈ Edit the pin number as desired.

◈ Click the Next Pin button to see the next pin in the list.

Auto-Numbering Features

An auto-numbering feature is provided to simplify numbering of sequential pins. If the Windows (CTRL) key or the Macintosh (OPTION) key is held down while a pin is clicked with the Text pointer, a new number is generated automatically for this item. The new label will be the same as the last one entered, except that the numeric part of the character string will have been incremented. If the previously-entered item did not have a numeric part, then a "1" digit will be appended to it. If the (SHIFT) key is pressed at the same time, the number will be decremented instead of incremented.

Setting Pin Number Text Style

The text style for pin numbers is set globally for the entire design. It cannot be set individually for pins.

To set pin number text style:

◈ Select the Design Preferences command in the Schematic menu.

◈ Click on the Pin Text button.

◈ Select the desired text font, style and size in the Font dialog.

◈ Click OK on the Font dialog, then OK in the Design Preferences dialog.

Depending on the size of the design, there may be a short delay at this point while sizes and positions of text items are recalculated.

Text Objects

Free text objects are used only to enhance the graphical appearance of a schematic diagram. They have no logical significance in the design.

IMPORTANT: Free text items are not associated with any particular device or signal on the screen, and should not be used to set a name or attributes for devices or signals. The text in these boxes is not accessible in net or component lists. Use the naming and attribute features to attach text to devices and signals.

Creating a Text Notation

If you click the text pointer on the diagram away from a device or signal line, a blinking cursor will appear at that point, and you will be able to type any desired text on the diagram. The Windows (ENTER) key or the Macintosh (RETURN) key can be used to enter multiple lines in a single text block. In the Windows version, text entry is terminated by clicking outside of the text entry box. In the Macintosh version, text entry is terminated with the (ENTER) key.

Editing Free Text

Windows—If you click the text pointer inside an existing text item, the insertion point will be positioned at the click point. You can then use normal text editing techniques to modify the text. Note that text on the Clipboard can be pasted into an existing text box using the (CTRL)–V key equivalent for the Paste function. The Paste menu command will cause the current text entry to be terminated and a new text box to be created. Similarly, the (CTRL)–key equivalents for Cut ((CTRL)–X) and Copy ((CTRL)–C) can also be used while editing a text box.

Macintosh—If you click the text cursor inside an existing text item the insertion point will be positioned at the click point. You can then use normal Macintosh text editing techniques to modify the text. Note that text on the Clipboard can be pasted into an existing text box using the ⌘–V key equivalent for the Paste function. The Paste menu command will cause the current text entry to be terminated and a new text box to be created. Similarly, the ⌘–key equivalents for Cut (⌘–X) and Copy (⌘–C) can also be used while editing a text box.

Text boxes can be zapped, duplicated, cut, copied, pasted, and dragged just like any other item on the screen. See the descriptions of these commands for more information.

Text Style and Display Options

To set text display options and text style, select the free text block by clicking on it with the arrow cursor, then select the Get Info command in the Schematic menu. This will display the following dialog:

Windows—

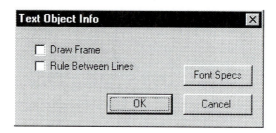

The following table summarizes the options available in this dialog.

Rule Between Lines	Turning this switch on causes a line to be drawn after each row of characters.
Draw Frame	Turning this switch on causes a frame to be drawn around the text item on the schematic.
Font Specs	Clicking this button displays the standard Font dialog. Any changes made in font style affect only the selected item, but they also become the default for future free text blocks.

Macintosh—

```
┌─────────────────────────────────┐
│  ┌───────────────────────────┐  │
│  │ Text Options              │  │
│  │    ☐ Draw Frame           │  │
│  │    ☐ Rule Between Lines   │  │
│  │     ┌─────────────────┐   │  │
│  │     │  Text Specs...  │   │  │
│  │     └─────────────────┘   │  │
│  │  ┌────────┐  ┌─────────┐  │  │
│  │  │ Cancel │  │   OK    │  │  │
│  │  └────────┘  └─────────┘  │  │
│  └───────────────────────────┘  │
└─────────────────────────────────┘
```

The following table summarizes the options available in this dialog.

Draw Frame   Turning this switch on causes a frame to be drawn around the text item on the schematic.

Rule Between Lines   Turning this switch on causes a line to be drawn after each row of characters.

Text Specs   Clicking this button displays the standard text style dialog. Any changes made in text style affect only the selected item, but they also become the default for future text blocks.

Sheet Borders and Title Blocks

LogicWorks provides a number of features to assist in creating the borders and title blocks required for a finished schematic diagram.

Creating a Sheet Border

Two methods are available for displaying or printing a border on the drawing:

■ The default border mechanism displays and/or prints a border with background grid lines and reference letters and digits at the edges. The border resizes automatically to match the current drawing size. This grid can be turned on and off using the options in the Design Preferences command.

■   To get more control over the appearance of the border, you can create a graphic of the desired size in any Windows application that will export Windows Metafile Format (WMF) data on the clipboard. This image can then be pasted onto the sheet and set to be a background object, using the procedure outlined below. This border will then be a fixed size and will not resize automatically with printer setup and drawing size changes. Any changes will have to be made manually to the original graphic which will then have to be re-pasted into the drawing.

Pasting Graphics onto the Diagram

Graphics from a number of sources can be pasted directly onto a Logic-Works schematic diagram:

■   Windows Metafile Format (WMF) data is exported by Microsoft Word and many drawing programs, and provides a clean, compact (i.e. a minimal amount of memory is used) and scalable image (i.e. prints cleanly on various types of printers). This is the recommended way of creating border and title block graphics.

■   Bitmap (BMP) images can be created using Windows Paint or many third-party paint programs. NOTE: BMP images are not suitable for large borders since they occupy a large amount of memory space and do not scale well when printing.

■   Graphics can be copied and pasted from the device symbol editor built into LogicWorks. This is a convenient way of creating images that do not require exact measures or sophisticated drawing tools. To do this, select the New command in the File menu, select the Device Symbol document type. Draw the desired graphics in the symbol editor, then Select All and Copy them onto the clipboard. Switch back to the schematic sheet and Paste the graphics onto the sheet. You can now close the device symbol editor without saving.

NOTE: There is an important difference between graphics created in the Device Editor using the above procedure and device symbols created in the Device Editor and then saved in a library and placed on the sheet from the library. When you copy and paste directly onto the sheet, you are creating only a graphic object, which has no circuit properties and no simulation and will not appear in any component lists. If you create exactly the same graphic, save it as a component in a library and then place it on the diagram, this will have an identical visual appearance, but will be treated within the program as a device. This means it will appear in component lists as a device and it can be given attributes, simulation parameters, etc.

Setting Graphic Item Properties

To set the properties of a graphical item on the diagram:

- Click on it once to select it (if the object has been previously set to be a background item, you will have to hold the (CTRL) and (SHIFT) keys (Windows) or ⌘ and (OPTION) keys (Macintosh) in order to select it.

- Select the Get Info command in the Options menu.

- Select the Draw Frame item to draw a border around the graphic.

- Select the Make Background item to prevent the item from being selected by a normal mouse click. Note the key sequence given above that is required to select a background object.

6
Advanced Schematic Editing

This chapter provides information on the more advanced schematic editing features of LogicWorks.

Bussing

The bussing facility allows any combination of named signals to be represented by a single line and any subset of these to be brought out through a "breakout" at any point along the bus line.

Properties of Busses

A bus is treated by LogicWorks as a signal with special properties. Thus, bus lines can be drawn and modified on the screen using all the same editing features available for signals. Note the following properties of busses:

- Only bus pins on devices can be connected directly to a bus. All other connections must be made by using a breakout to access the desired internal signals. A breakout is created using the New Breakout command in the Schematic menu.

- You do not need to specify in advance what signals will be contained in a given bus. Any signals that are present in a breakout or bus pin attached to a bus will become part of that bus and can be brought out through another breakout anywhere along the bus.

- Any two busses can be joined together, regardless of their internal signals. When two different busses are merged, any signal in either bus becomes available anywhere along the combined bus.

■ If you select a bus line, then pull down the Schematic menu and select the Get Info command. The displayed info box will show a list of the signals currently contained in the bus.

■ A given signal can be present only in one bus. If you attempt to connect together two signals in different busses, a warning box will be displayed and the connection will be canceled.

■ A bus can be created by drawing the bus lines first, then creating the breakouts to attach, or by creating a breakout and extending the bus line starting at the bus pin. Bus lines are drawn or extended using exactly the same techniques as for signals, except that the Draw Bus command or cursor is used instead of Draw Signal.

Properties of Breakouts

Signals are attached to a bus via a special type of device symbol called a "breakout." It is not legal to attach a signal line directly to a bus line. If a signal line touches a bus line, no connection will be made. In LogicWorks, a breakout is treated as a device with certain special properties. This means that it can be placed in any desired orientation, moved, duplicated, etc., using any of the device editing features available. A typical breakout appears as follows:

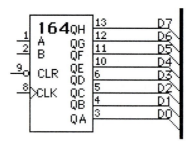

Any breakout can always be attached to any bus. When a breakout is attached that contains signals unknown in that bus, the signals are implicitly added to the bus. For example, suppose we want to add control signals to the above circuit. We could create a breakout containing only the new signals, as follows:

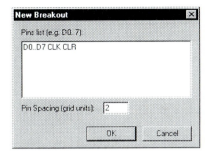

Once such a breakout has been added to the bus, all signals in all attached breakouts are considered part of that bus. A list of internal signals can be seen by selecting the bus and using the Get Info command:

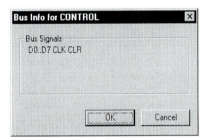

Any combination of the internal signals can now be brought out of the bus at any point, as in the following addition to the above circuit:

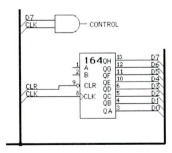

Bus Operations

Creating a Bus

A bus can be created by any one of the following methods:

▨ Select the Draw Bus tool (**+**) in the Tool Palette. Draw any desired contiguous set of lines on the diagram using the usual signal drawing techniques. This bus will have no internal signals initially. Signals will be added implicitly when it is connected to any breakout or bus pin.

▨ Create a breakout symbol using the New Breakout command (see below). The bus pin (backbone) of the breakout can now be extended using the normal pointer (**↖**) or the Draw Bus cursor. The bus will contain all signals specified in the breakout.

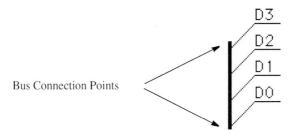

Bus Connection Points

▨ Extend a line out from an existing bus pin on a device (see below) using the normal pointer or the Draw Bus cursor. The bus will contain all signals specified in the bus pin on the device. Connections between bus internal pins and bus internal signals can be changed using the Bus Pin Info command on the bus pin's pop–up menu.

Adding Signals to a Bus

There is no explicit command to add signals to a bus. Signals are added to a bus each time a breakout or device bus pin is connected to the bus. Any signals in the breakout or bus pin are implicitly added to the bus if they don't exist already.

Creating a Breakout

To create a breakout, select the New Breakout command in the Schematic menu. If the new breakout is to be similar to an existing one, first select the similar breakout or the bus to which the new breakout is to be connected. Then select the New Breakout command. The following dialog box will appear:

If a bus or breakout was selected on the circuit diagram, the New Breakout Info dialog will display a list of the signals in that bus or breakout; otherwise, it will be empty. If this list already matches the signals you want in the new breakout, then just click the "OK" button or press (ENTER) on the keyboard. Otherwise, edit the signal list, noting the following options:

▨ Blanks or commas can be used to separate individual names in this list; therefore bussed signals *cannot* have names containing a blank or comma.

▨ A range of numbered signals can be specified using the following formats:

D0..7 or D0..D7

is equivalent to

D0 D1 D2 D3 D4 D5 D6 D7

D15..0

is equivalent to

D15 D14 D13 D12 D11 D10 D9 D8 É D0

D15..D00

is equivalent to

D15 D14 D13 D12 D11 D10 D09 D08 D07 É D00

Note that the ".." format implies that bussed signal names cannot contain periods.

■ The signals specified will always appear in the order given in this list from top to bottom in standard orientation. Specifying numbered signals from lowest numbered to highest is a good practice, as in the first example above, since this matches the standard library symbols.

■ There is no fixed limit on the number of signals in a bus, but it is a good practice to divide busses up by function (that is, address, data, control, etc.) for ease of editing.

■ Any combination of randomly–named signals can be included in the list, as in the following examples:

D0..15 AS[*] UDS[*] LDS[*]

CLK FC0..3 MEMOP BRQ0..2

Once the list has been entered, click on the OK button or press the (ENTER) key. A flickering image of the breakout will now follow your mouse movements and can be placed and connected just like any other type of device.

Editing Breakout Pins

The signal name notation that appears on a breakout pin is actually a pin attribute. It can therefore be edited by the usual attribute editing mechanisms—that is, either:

◆ Select the pin and choose the Get Info command in the Schematic menu, then click the Attributes button;

or:

◆ Click the text cursor directly in the text on the schematic, as illustrated:

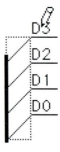

◆   Type the desired new name.

◆   Press the (ENTER) key. The breakout pin and the attached signal will be renamed as entered.

IMPORTANT: The notation on the breakout pin is always the same as the name of the attached signal. Changing the breakout pin renames the attached signal and will detach it from any like–named signals already in the bus.

Changing Bus Pin Connections

When a bus is connected to a bus pin on a device or subcircuit block, the bus internal pins will by default connect to signals with the same name in the bus. To change these default connections, use the Bus Pin Info command in the pin pop–up menu.

◆   See Chapter 12, Menu Reference, for more information.

Bus Pins

LogicWorks supports user–created bus pins on devices. A bus pin can be defined to have any collection of named internal pins. Note the following properties of bus pins:

▩   The bus pin itself does not represent a physical device pin. It is only a graphical place–holder on the schematic representing a group of internal pins. The bus pin itself never appears in a netlist.

▩   The internal pins represent physical device pins. Even though they do not appear on the schematic, they can have all the same parameters as normal devices pins, including pin numbers and attributes. These parameters can be accessed using the Bus Pin Info command in the pin pop–up menu.

▩   When a device with a bus pin is placed, it has a pre–created bus attached to it by default. This bus will contain one signal for each internal pin, with the initial name of the signal being the same as the name as the pin's name.

▨ A "splicing" box can be displayed using the Bus Pin Info command in the pin pop–up menu. This box allows any internal pin to be connected to any signal in the attached bus.

◆ For more information on creating device symbols with bus pins, see Chapter 10, Device Symbol Editing.

Power and Ground Connections

LogicWorks uses a type of pseudo–device symbol called a "Signal Connector Device" to maintain connectivity between like–named power and ground symbols that are used on circuit diagrams.

As soon as a Ground symbol is placed on the diagram, the attached signal will be named "Ground" (the name will initially be invisible). This will cause it to be connected by name to any other signals that have Ground symbols or are explicitly named "Ground".

Connectivity can be checked at any time by double–clicking on any ground or power line. This will highlight all other like–named lines on the diagram.

IMPORTANT: Signal connectors *do not* cause a logical connection to be made between circuit levels in nested subcircuits.

Using Signal Connector Devices

Signal Connector Devices are placed on the diagram just like any other LogicWorks device. A set of standard power–supply symbols are included with LogicWorks in the connectors or pseudo devices libraries.. If you connect two different signal connector devices together, you will be prompted to provide a name for the resulting signal.

Creating Signal Connectors in a Library

Signal Connector devices are special primitive "pseudo–devices" in Logic-Works and can be created using the Set Primitive Type command in the device symbol editor to select the SIGCONN primitive type.

IMPORTANT: The signal attached to a signal connector device is actually named to match the *pin name* of the signal connector pin specified in device symbol editor, *not* the type name. In most of the power and ground symbols provided with LogicWorks, these two names are the same. However, it is possible to create a symbol called "Ground" (for example) in a library that actually names the attached signal "GND". The Ground symbol in the spice.cct library is an example of this—it names the attached signal "0" to match the SPICE ground–naming convention.

◆   See Chapter 11, Device Symbol Editing, for more detailed information on this procedure.

Connectors and Discretes

In LogicWorks, each symbol is considered to be a separate device and each device is normally assumed to be one IC package with a standard pin numbering scheme. Thus connectors and discrete components will require special consideration.

Handling Connectors

Connectors can be handled in one of two ways:

▓   A special symbol can be created for the connector with the appropriate number of pins and pin numbering specified for each pin. This can be done using the device symbol editor to create a device symbol using your own picture.

▓   Each connector pin can be created as a separate single–pin device or as a custom symbol. The second option is preferable only if you need to spread the connector pins over different parts of the diagram. In this case, each "device" must be given the name of the connector and the pin number associated with that pin. The report generator will normally merge all devices with the same name into a single component entry.

Following is an example of these two methods:

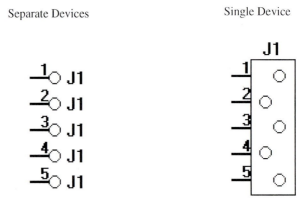

Separate Devices   Single Device

NOTE: When the single–pin devices are used, every device must carry exactly the same name, although the names can be invisible if desired.

Handling Discrete Components

Discrete components—such as capacitors, transistors, etc.—can be handled just like any other device, except for the following special considerations.

Pin Numbering on Discrete Components

Pin numbers are not normally placed on discrete component pins on a diagram. If pin numbers are omitted from a device, LogicWorks will normally put a question mark in the netlist item for that device. Two methods are available to provide pin numbers for netlisting purposes:

■   To provide automatic numbering of discrete devices pins, the Report Generator provides an auto–numbering option. This option causes any device with less than or equal to three pins to be numbered automatically if no pin numbers are present on the diagram.

IMPORTANT: This option assumes that the pin number order of the discrete components is not significant. If a specific order is important, do not use this method.

* Pin numbers can be assigned but left invisible. This is done using the Get Info command for either the pin or the device.

Using Attributes

LogicWorks allows arbitrary blocks of text to be associated with any device, signal, or pin in a design, or with the design itself. The blocks of text are called *attributes*. Attributes have a wide variety of uses, including:

* Displaying device name, component value, etc.
* Storing data for use by external systems such as simulators, PCB layout, analysis tools, etc.

Default Values

A device symbol can incorporate predefined default values for any number of fields. Values can be specified for the device itself, and independently for each pin on the device.

When the standard Attributes Dialog is displayed for a device, you will see a button labeled Use Default Value. If this button is grayed out, then there is no default value, or the value shown is already the default.

◆ See Chapter 11, Device Symbol Editing, for more information on creating default attribute values.

Attribute Limitations

Attribute fields have the following specific limitations:

* Length of field name: 16 characters
* Length of field data item: 32,000 characters.
* Number of displayed positions of a single attribute item: 100

Like all other circuit data, the amount of attribute data that can be associated with a design is limited by available memory.

Predefined Attribute Fields

The following table describes the fixed list of attribute fields provided in each LogicWorks design. Attribute fields cannot be added or deleted in LogicWorks.

Field Name	Used In	Description
CctName	Design	Design file name. Sets the window title and name of next saved file.
Delay.Dev	Device	Specifies device delay. For most devices, a single decimal integer 0 to 32,767. For Clock and One Shot devices, two integers separated by commas. Should be set using the Parameters command, and not edited manually.
Delay.Pin	Pin	A decimal integer specifying pin delay in the range 0 to 32,767. Should be set using the Windows Parameters command or the Macintosh Simulation Params command and not edited manually.
Initial.Pin	Pin	This field is used to specify the initial state for storage devices when a Reset or Clear Simulation operation is performed. It can contain a single character, either 0, 1, X, or Z.
Initial.Sig	Sig	This field is used to specify the initial state for a signal. It can contain a single character, either 0, 1, X, or Z.
Invert.Pin	Pin	This field is used to specify logical inversion on device pins. Any non–empty value indicates inversion should be done.
Name	Device, signal	The device or signal name. This is the field set using the text tool on the schematic or the Name command in the pop–up menu.
Spice	Device, Design	Holds simulation parameters for SPICE–based simulators. Not used internally.
Value	Device	Component value to appear on the schematic. Not used internally.

Editing Attribute Data (General)

The following dialog box is used to enter or edit attribute data:

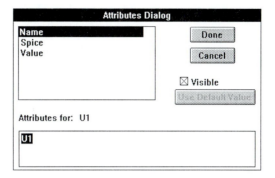

NOTE: The same Attributes Dialog is used to enter data for all object types. This section discusses the general operation of this dialog. The following sections will discuss each object type.

Basic Procedure

To edit the contents of a field, simply select the field name in the list. The current contents of the field will be displayed in the editable text box. Edit this value using the normal text editing techniques. Select another field or press the Done button if you are finished editing. If the data you typed exceeded the maximum length for the field, or if it contained invalid characters for the field, then you will be asked to correct the data.

You can view or edit as many fields as desired while in this dialog. No changes are made to the actual design data until you click the Done button. Clicking Cancel will abandon all changes made while in this dialog.

Default Value

Clicking the Use Default Value button sets the value for the selected field to the default value stored with the symbol. If this button is inactive (grayed out) then the value is already the default value, or no default value is present. Only devices and pins can have default values.

Editing Device Attribute Data

The Attributes Dialog can be entered in one of two ways:

▓ Click on the device to select it, then select the Get Info command from the Schematic menu, then click on the Attributes... button. Or:

▓ Display the device's pop–up menu (Windows users right–click on the device; Macintosh users ⌘–click on the device). Then select the Attributes command from the menu.

The standard Attributes Dialog will appear. Select the desired field by clicking on it in the list. The current contents of the selected field will be displayed in the text edit box. This text may be edited using standard editing techniques.

Displaying an Attribute on the Schematic

To display device, signal, or pin attribute text on a schematic:

◆ Display the pop–up menu for the device, signal, or pin to which you want to attach the attribute. (Windows users right–click on the device; Macintosh users ⌘–click on the device.)

◆ From the pop–up menu, select the Attributes command to display the Attributes Dialog.

◆ Select the desired field by clicking on its name in the field list.

◆ Edit the attribute value as desired.

◆ Turn on the Visible switch.

◆ Click OK.

The attribute text will now be displayed in a default position near the device or signal. It can be dragged to any desired location using the Point tool.

Rotating Attribute Text

To rotate an attribute text item that is already displayed on the schematic:

◆ Display the attribute pop–up menu for the text item you want to rotate. (Windows users right–click on the text item; Macintosh users ⌘–click on the text item.)

◆ From the pop–up menu, select the Rotate Right or Rotate Left command.

Setting Attribute Text Style

Attribute text style is set globally for the entire design. There is no way to set text style for an individual item.

IMPORTANT: Changing the attribute text style affects all visible attributes throughout the design. LogicWorks may alter text alignment and position to accommodate a new text size.

To set the global text style:

◆ Choose the Design Preferences command in the Schematic menu.

◆ Click on the Attr Text... button.

◆ Select the desired font, style and size, then click OK.

◆ Click OK in the Design Preferences dialog.

Depending on the size of the design, there may be some delay at this point. The program must check all visible attribute items to see if their position and framing is affected by the text change.

Using Subcircuits

LogicWorks provides the ability to have a device symbol in a schematic actually represent an arbitrary circuit block. This subcircuit can be used to implement a simulation model for a device of arbitrary complexity. Subcircuits can be nested to any desired depth, so devices containing subcircuits can themselves be used as subcircuits for more complex devices. For clarity, a device symbol that represents an internal circuit will be called a "subcircuit device" in the following text.

Following is a short summary of the rules governing subcircuit devices. More information on each of these topics is included in the following sections.

- The "pins" on the subcircuit device symbol represent connections to specific input–output points on the internal circuit. A "port connector" pseudo–device must be placed in the subcircuit corresponding to each pin on the parent symbol. Port connector symbols are found in the connect.clf library supplied with LogicWorks.

- A subcircuit device can be opened at any time by double–clicking on the parent symbol. Subcircuits can be "locked" to prevent accidental modification by selecting the Lock Opening Subcircuit option in the Device Info box.

- Subcircuits cannot be "recursive," i.e., you cannot use a device symbol inside its own internal circuit.

- The netlist and bill of materials reports generated by the Report tool in LogicWorks only list components in the top–level circuit in the design. Devices in subcircuits are never listed.

- A device symbol with an associated subcircuit can be stored in a part library. Each time that symbol is selected from the library, the subcircuit definition will be loaded and attached to the device.

- When you open a device's subcircuit, a temporary copy of the subcircuit is made to isolate it from all others of the same type that have been used elsewhere in the design. When you closed the subcircuit, choosing the "update" option will cause all other devices of the same type to be modified.

- If a given type of subcircuit device has been used more than once in the same design, you can only have one of them open at a time for viewing or editing the subcircuit.

- Signals in an *open* subcircuit can be displayed in the Timing window. As soon as the subcircuit is closed, the waveforms for any of its signals that were displayed will be removed.

A Simple Subcircuit Example

The following diagram is the master circuit, or top level, of our design example:

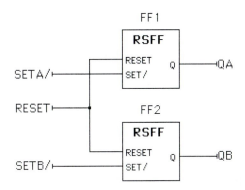

Note that it contains two symbols, both representing subcircuit devices. Both symbols are of the same type, RSFF, and therefore share the same internal circuit definition. The two devices are named FF1 and FF2. Opening either one of these devices reveals the following internal circuit:

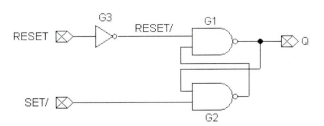

This circuit consists of three device symbols, G1, G2, and G3, representing physical devices, and a number of port connector symbols. The port connectors define the interface between the internal circuit and the pins on the symbol representing it.

Note the following characteristics of this simple design:

▓ The device RSFF has been used twice, so there are actually *two* G1s, one inside FF1 and one inside FF2. We say that there are two *instances* of G1. Similarly for G2 and G3.

■ The signals SET/, RESET, and Q in the internal circuit will actually get absorbed into the attached signals in the parent circuit because they are attached to port connectors. They do not exist independently in the physical circuit.

■ The signal RESET/ does not connect to a port connector, so it represents a separate signal in the internal circuit. Like the devices G1, etc., each signal in RSFF actually represents two physical signals.

Subcircuit Primitive Type

Subcircuit device symbols are simply device symbols which have the primitive type "SUBCCT." Device symbols with any other primitive type cannot be used as subcircuit devices. SUBCCT is the default primitive type when creating symbols with device symbol editor, so it is normally not necessary to change this setting.

Port Interface

Signal connections between circuit levels are made using port connector symbols. With the exception of power and ground nets, all connections between levels must pass through a port connector.

Port/Pin Naming

The relationship between the port connector in the subcircuit and the pin on the parent device symbol is established by matching the pin name on the parent device with the Name field of the port connector. For example, if we were to open the RSFF device used in the example above using the device symbol editor, we would see the following pins listed:

For a complete port interface, a port connector must exist in the internal circuit named to match each one of these pins. In this case, the following port connectors would be required (ignoring all other internal circuitry):

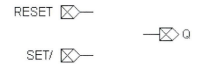

The port interface is rechecked whenever any change is made. Thus, as soon as a port connector is added or removed, or its name is changed, the port interface will be updated to reflect the new logical connections. However, to avoid excessive warning messages, error checking is performed only when an internal circuit is opened or closed. A warning box will be displayed if any error is found. This checking cannot be disabled.

NOTE: The name of the port connector's pin and the name of the signal attached to the port connector are *not significant* in making the port association. Only the contents of the port connector's Name field are used. Note the different rules for bus ports below.

Port Pin Type

In order for the simulation of a subcircuit device to operate correctly, the type of port connector symbol used in the subcircuit must match the type of pin on the parent device symbol, according to the following table:

Parent Pin Type	Port Connector (in the connect.clf library)
Input	Port In
Output	Port Out
Bidirectional	Port Bidir
Bus	Must be custom-made
All others	Port In[*]

[*] For Tied High, No Connect, and other pin types, use a Port In for consistency —although no simulation data is transferred through these types of pins in any case.

IMPORTANT: If you create a port connector symbol using the device symbol editor, the pin type (input, output or bidirectional) must be set carefully for each pin on the port connector. The pin on the Port Connector symbol must be of the *opposite* type to the corresponding pin on the parent device symbol. For example, a signal coming *in* to the subcircuit is actually an *output* from the port connector pin.

Note, for example, that the pin on the Port In device in the connect.clf library is set to be an output and the Port Out device has an input pin. A bidirectional port has bidirectional pins on both sides of the interface.

Bus Ports

Connections can be made between busses across circuit levels using Bus Port Connectors. Bus pins on a parent device symbol must be matched with a Bus Port Connector having identical internal pins. For this reason, Bus Port Connectors *must always be custom–made* using the device symbol editor.

Bus Pin Name Matching

Note the following rules for name matching in bus ports:

- As with other Port Connectors, a Bus Port Connector must be given a name exactly matching the pin name of the *bus pin* on the parent device.
- The internal pins in the parent bus pin must exactly match the internal pins on the Bus Port Connectors bus pin.
- The pin name of the bus pin itself on the Bus Port Connector is not significant.
- As with normal ports, the names of the signals attached to the Bus Port Connector's pin are not significant.

Bus Pin Example

For example, the following simple device has a bus pin called CONTROL containing internal pins CLK, MEMW/, and MEMR/.

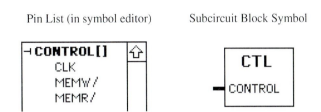

Pin List (in symbol editor)   Subcircuit Block Symbol

The corresponding Bus Port Connector to be used inside this device would look as follows:

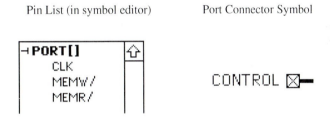

Pin List (in symbol editor)   Port Connector Symbol

The comments above in the section, Port Pin Type, apply to each internal pin in a bus pin. Remember that the name of the bus pin in the port connector is not significant.

Power and Ground Connections

Power and Ground symbols (for example, signal connector devices) *do not* make a logical connection across subcircuit levels. For this reason, signal connectors should not be used to make active signal connections for interactive simulation purposes.

They can be used to tie signals to high or low values, however, since it is not relevant whether all tied–high signals are actually interconnected.

Creating a Subcircuit—Top–Down

To create a subcircuit top–down (for example, creating the subcircuit itself after the parent symbol has already been used in a circuit), follow these steps:

◆ Create the parent symbol using the device symbol editor. Be sure to set the pin type (in/out/bidirectional) appropriately for each pin on the symbol. (The Subcircuit / Part Type command does not normally need to be used because the default primitive type for a symbol is SUBCCT.) Save the symbol in a library.

◆ Use the symbol as desired in your schematic.

◆ Use the New Design command to create a new and completely independent design. Create the schematic for the subcircuit in this design. You may use any existing parts from libraries *except* the parent symbol that we created above. Subcircuits cannot be recursive!

◆ Add port connectors to the design and attach them to the appropriate connection points. Each port connector must match its corresponding pin in type according to the following table:

Parent Pin Type	**Port Connector (in the connect.clf library)**
Input	Port In
Output	Port Out
Bidirectional	Port Bidir
Bus	Must be custom–made
All others	Port In [*]

[*] For Tied High, No Connect, and other pin types, use a Port In for consistency, although no simulation data is transferred through these types of pins in any case.

◆ Name each port connector to match the associated parent pin. You may want to have the parent symbol open in the device symbol editor at the same time so that the names are easily checked.

NOTE: There *must* be a one–to–one match between the pins on the parent symbol and the port connectors in the subcircuit.

◈ Return to the design where the parent symbol was used. Select the parent symbol by clicking on it. If it has been used more than once, select any of the copies.

◈ Select the Attach Sub-Circuit command from the Schematic menu. Choose the design containing the subcircuit from the list of open designs, then click the Attach button. The selected design will now be brought to the front. Close its window. If the Update/Revert/Cancel option box appears, select Update.

The subcircuit is now attached to the parent symbol and has ceased to exist as an independent design.

Creating a Subcircuit—Bottom–Up

In a bottom–up design process, we create the subcircuit first then use it to define the pins on the parent symbol. In LogicWorks, this is easier than the top–down procedure because we can take advantage of some of the automatic features of the device symbol editor for this purpose. The bottom–up procedure is as follows:

◈ If you are creating the subcircuit from scratch, select the New Design command from the LogicWorks menu bar to create a new circuit window, then use the schematic drawing tools to draw the circuit. You may use any existing parts in creating the subcircuit, including other subcircuit devices. Alternatively, if the subcircuit is to be based on an existing circuit file, open that file using the Open Design command.

◈ If you haven't already done so, add Port Connectors corresponding to the pin connections on the symbol, as described in the previous section.

◈ Leave this circuit open (that is, displayed in a circuit window). You may save this circuit to a file if desired, but it is not necessary to perform this procedure.

◈ Open the device symbol editor. On the Macintosh, select the DevEditor item from the LogicWorks menu bar's Tools menu. On Windows, select New in the File menu and then choose the Device Symbol option. From the Options menu (Macintosh: DevEdit menu), select the Subcircuit / Part Type command and choose the "Create a subcircuit symbol and store the subcircuit with it..." option. Select the subcircuit that you just created from the list of open windows that is presented. Close the PartType configuration dialog. You will notice that symbol editor has extracted the names from the port connectors in the subcircuit and placed them in the Pin List at the left side of its own window.

◆ Create the graphics for the symbol using either drawing tools, the Windows Autocreate Symbol command on the Options menu, or the Macintosh Auto Create Symbol command. Every pin listed in the Pin List must have a corresponding graphical pin on the device symbol.

◆ Save the symbol to the library. It will be saved with a *copy* of the selected internal circuit; that is, you can close or discard the internal circuit window, as the circuit is now saved in the library.

The new subcircuit device may be selected from the library and placed in any schematic as desired.

◆ For more information on associating a subcircuit with a part in a library, see the section, Creating a Part with Subcircuit, in Chapter 11, Device Symbol Editing.

7
Simulation

This chapter provides more detailed information on LogicWorks'
simulation capabilities.

General Information on Simulation

LogicWorks has the ability to perform a realistic simulation of any digital
circuit. Obviously, though, any simulation of any system must be limited in
detail and must make certain assumptions. In particular, when simulating
digital circuits, it must be understood that real circuits are never completely
"digital" in nature, and that they in fact have many "analog" properties
which affect how they operate.

LogicWorks is primarily intended to assist with the logical design of a circuit, and does not take into account factors such as line loading, power supply noise, rise and fall times, output drive, and so on. As more of these
factors are taken into account, the simulation becomes slower and less
interactive, which defeats the purpose for which LogicWorks was created.

Type of Simulation

LogicWorks performs a discrete simulation of the signal changes in a logic
circuit, meaning that signal levels and time change only in steps, rather
than continuously. The program does not attempt to analyze your circuit,
but simply tracks signal–level changes through the devices. Thus, circuits
with feedback loops or other delay–dependent features will be simulated
correctly as long as they don't rely on particular analog characteristics of
devices.

The simulation is "event–driven," where an event is a change in the level of a signal. Each time an event occurs, a list is made of all the devices whose inputs are affected by that event. Any other events occurring at the same time are similarly evaluated, and affected devices added to the list. A type–specific routine is then called for each device on the change list in order to determine what output changes are going to occur. These changes are added to the event list, their time of occurrence depending upon the device delay. No computation is performed for times when no event occurs—so that device delay settings and clock values have no effect on how fast the simulation is performed.

LogicWorks performs strictly a digital simulation. It does not take into account factors such as fan–out (that is, the number of inputs connected to a given output), line length (capacitance), asymmetrical output drive, and so on, except inasmuch as these affect delay time.

Simulation Memory Usage

When a circuit is opened or created by LogicWorks, the circuit data is retained completely in the memory of your machine. Since the total memory available is fixed (until you buy your next memory expansion!), this places some limits on circuit size and simulation.

Each time a signal changes state, an "event" record is created in memory. If the signal is not being displayed in the Timing window, this record is deallocated again after the signal change has occurred. If the signal is being displayed, then the record is retained in memory until that change has scrolled off the left–hand side of the Timing window. As a result, the memory used by event records will increase when the number of displayed signals is increased or the resolution of the timing display is decreased. Memory usage will also increase if the "retain time" setting is increased.

Time Units

LogicWorks uses 32–bit signed integer arithmetic to calculate all time values used in the simulation. It is usually convenient to think of these values as being in nanoseconds, but the actual interpretation is left up to the user.

The simulation will stop if any time value approaches the 32–bit integer limit.

Signal Simulation Characteristics

Signal States

LogicWorks uses 13 different device output states in order to track conditions within your circuit. These states can be broken into three groups, as follows:

Forcing States (denoted by suffix .F):

> LOW.F
>
> HIGH.F
>
> DONT01.F
>
> DONT0Z.F
>
> DONT1Z.F
>
> CONF.F

Resistive States (denoted by suffix .R):

> LOW.R
>
> HIGH.R
>
> DONT01.R
>
> DONT0Z.R
>
> DONT1Z.R
>
> CONF.R

High Impedance:

> HIGHZ

Note that the Forcing/Resistive distinction is used only to resolve conflicts between multiple outputs connected to the same signal. The final value stored or displayed for a given signal line can only be one of five possibilities:

> LOW
>
> HIGH
>
> DONT
>
> CONF
>
> HIGHZ

Description of States

The High and Low states are the normal ones expected in a binary circuit, but are not sufficient to realistically simulate circuit operation, so the High Impedance, Don't Know and Conflict states are added. There will always be some cases where the simulation will not correctly mimic what would appear in a real circuit, and some of these cases are discussed in following sections. In particular, if a circuit takes advantage of some analog property of a specific device—such as inputs that float high, known state at power–up, input hysteresis, and so on—it is unlikely to simulate correctly.

High Impedance

This state ("Z" on a logic probe) is used for cases when no device output is driving a given signal line. This may occur for an unconnected input, or for a disabled "three–state" or "open–collector" type device. If a device input is in the High Impedance state, it is treated as unknown for the purposes of simulation, even though in a real circuit the device may assume a high or low state, depending on the circuit technology used.

Don't Know

The Don't Know state ("X" on a logic probe) results when the simulator cannot determine the output of a device. This may occur, for example, when an input is unconnected or when the output from a previous device is unknown. The Don't Know signal will be propagated though the circuit, showing the potential effects of that condition.

The Don't Know state is used in LogicWorks in cases where the actual result in a real circuit would depend on the circuit technology used, on random chance, or on analog properties of the device not predictable using a strictly digital simulation. For example, if the following ring oscillator circuit is created in LogicWorks, all signals will be permanently unknown—since each depends on the previous one, which is also unknown. In actual hardware, this circuit may oscillate, or may settle into an intermediate logic level, which would not be defined in a digital circuit.

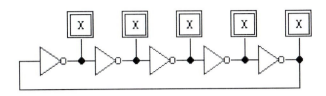

For the purposes of simulation, all circuits must have some provision for initialization to a known state. In most cases, circuits can be initialized by using the Clear Unknowns command or by setting the initial value attribute, described in "Setting Initial Values" on page 165. Alternatively, circuitry can be added to allow a reset to be done, as in the following modification to the ring oscillator:

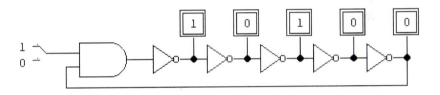

A problem arises in simulating circuits with multiple open collector devices—such as a bus line, illustrated here:

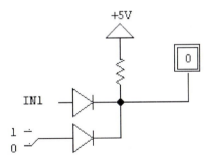

In this circuit, the upper device has an unconnected input at IN1 and there-fore outputs a Don't Know value. The lower device has a low input and therefore outputs a low value. In order to correctly resolve this situation the simulator needs to distinguish between a Don't Know output from a normal "totem–pole" type output and a Don't Know from an open–collector, open–drain, or other single–drive output. In this case, the upper device will produce a DONT0Z output, which resolves correctly to a LOW on the output—regardless of the state of IN1—using the rules described previously.

Conflict

The Conflict state ("C" on a logic probe) results when two device outputs are connected and are of different or unknown states—taking into account the rules described previously.

State Display

The Timing window displays the various signal states in different colors (Windows version) or in different patterns (Macintosh version).

The following Timing window shows how the various signal states are displayed.

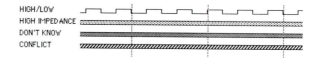

Stuck–At Levels

The LogicWorks simulator implements stuck–at levels to assist in setting initial simulation states, testing for faults, and so on. When a signal is in a stuck–at state, it will not change state, regardless of changes in devices driving the line.

When the stuck–at status is set, the signal will retain the value it had at that time—until some user action forces a change. When the stuck–at status is removed, the signal will return to the value determined by the devices driving the line.

Setting Stuck Levels

A signal can be placed in a Stuck–High or Stuck–Low state by any of the following means:

■   Applying the name "0" or "1" to the signal;

- Typing "H" or "L" while viewing the signal value with the signal probe tool; or,

- Using the Stick High or Stick Low buttons in the Stick Signals command.

Each of these methods is described in more detail in the relevant section of this manual.

Clearing Stuck Levels

The stuck status can only be cleared by one of the following user actions:

- Typing the spacebar while viewing the signal using the signal probe tool; or

- Clearing the "stuck" switch in the Stick Signals command.

Resolution of Multiple Device Outputs

The DONT0Z and DONT1Z values are used primarily to handle cases of open collector or open emitter devices with unknown inputs (see following additional information). Most other types of devices produce the DONT01 output when a value cannot be calculated.

In cases where two or more device outputs are connected together and each one drives the line with a different value, the following rules are used to resolve the actual value on the line:

- The forcing/resistive distinction is only used to resolve outputs from multiple devices. The final value used for display and simulation purposes is one of the forcing values or HIGHZ.

- A forcing drive always overrides a resistive drive or HIGHZ (that is, the signal takes on the value of the forcing drive, ignoring all resistive drives and HIGHZs).

- A resistive drive always overrides HIGHZ.

- DONT0Z.F and LOW.F produce LOW.

- DONT1Z.F and HIGH.F produce HIGH.

- Any other combination of conflicting forcing drives produces CONF.

- DONT0Z.R and LOW.R produce LOW.

■ DONT1Z.R and HIGH.R produce HIGH.

■ Any other combination of conflicting resistive drives produces CONF.

Resistive vs. Forcing Drive

All primitive devices in LogicWorks output a forcing drive level, except for the Resistor primitive device. The function of the Resistor device is to convert a forcing drive on one side into a resistive drive on the other. This can be used to modify the output of any existing device type by placing a resistor in series with it. Note that LogicWorks does not model analog properties of devices, so the resistor does not have a resistance value in the analog sense. In particular, there is no interaction between resistor and capacitor symbols to produce delay in lines. The delay effect can be simulated by setting a delay value for the resistor.

Signal Probe Tool

The Signal Probe tool allows you to interactively examine *and* change values on individual signals and pins in the circuit diagram. When the probe tip is clicked and held on a signal line or pin, the cursor will show the current value on the signal or pin, and will track changes that occur as the simulation progresses.

Probing a Signal

Only the signal under the cursor at the time of the click is examined; moving the mouse while the button is pressed does not change the signal being viewed.

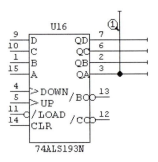

Probing a Pin

If the probe tip is clicked on a device pin close to the device body, the probe shows the driving level of that pin, rather than the state of the attached signal. This can be used to resolve drive conflicts in multiple drive situations, as in the following example using open collector buffers:

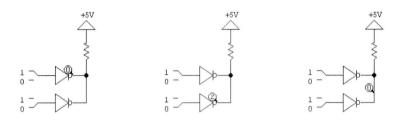

Pin Drive on Upper Device   Pin Drive on Lower Device   Combined Signal Value

NOTE: The probe display does not distinguish between low and high drive levels.

Injecting a Value Using the Probe Tool

While the mouse button is held, you can press keys on the keyboard to inject new values onto a signal, as follows:

0	LOW.F
1	HIGH.F
X	DONT01.F
C	CONF.F
Z	HIGHZ
L	LOW.F stuck
H	HIGH.F stuck
space	unstick

If a stuck value is forced onto a signal, the signal will not change state until the stuck value is cleared by some user action, regardless of device outputs driving the line. If a non–stuck value is forced, the signal value will revert to its appropriate new level when any change occurs on a device output driving the line.

The spacebar "unstick" command causes the signal to revert to its driven value.

◆ See also the Stick Signals command in Chapter 12, Menu Reference, for more information on stuck values.

Busses

Busses—that is, groups of signals represented by a single line on the schematic—have no particular significance to the simulator. The value of a bus is completely determined by the values of the individual signals it contains. The simulator performs no operations on the bus itself.

NOTE: You can display a bus in the Timing window using the Add To Timing command. This is equivalent to displaying all the internal signals individually and then grouping them.

Bus Pins

Bus pins, like busses, have no particular significance to the simulator. The value of a bus is completely determined by the values of the individual pins it contains. The simulator performs no operations on the bus pins themselves. Bus pins are not supported on primitive device types.

Device Simulation Characteristics

Device and Pin Delay

This section describes how to set delay values for primitive devices, subcircuit devices, and pins.

Primitive Device Delay

Primitive devices (e.g., those with a program–defined simulation model) have a single delay value which can be set to any integer value from 0 to 32,767. This delay is applied when any input change causes any output change. In addition, a pin delay in the range 0 to 32,767 can be set on any input or output pin. Pin delays can be used to set arbitrary path delays through the device. More information on pin delays follows.

The initial delay value is set to 1 when the device is created, but this can be changed later using the Parameters command on the Windows Simulator menu or the Simulation Params command on the Macintosh Simulate menu. This delay applies whenever any input change causes an output change. There is no provision in the built–in simulation models for different delay values on low–to–high and high–to–low transitions. The Clock and I/O devices have no delay characteristic. See the following notes on delay in subcircuit devices.

Subcircuit Device Delay

Subcircuit devices inherit their delay characteristics from their internal circuit and have no "device delay" characteristic of their own. The Parameters (Windows) or Simulation Params (Macintosh) command cannot be directly used on a subcircuit device, although pin delays *can* be set separately on each instance of a subcircuit device to customize path delays.

Pin Delays

Any input or output pin on any device (including port connectors and subcircuit devices) can have a pin delay associated with it. Pin delays normally default to 0 time units, but can be in the range 0 to 32,767.

A pin delay acts like a "buffer" device with the given delay inserted inline with the pin. On an input pin, the device simulation model will not see a change in signal value until after the pin delay has elapsed. On an output pin, the pin delay is added to the overall device delay for any changes scheduled on that pin.

Setting the Delay

To set the delay for a device, first select the device by clicking on it. Then choose Parameters from the Windows Simulator menu or Simulation Params from the Macintosh Simulate menu.

A dialog box will appear, allowing you to increase or decrease the delay value by clicking one of two buttons. The minimum delay value is 0 and the maximum is 32,767. When the delay setting for a subcircuit device is changed, the delays for all internal devices are changed by the same amount.

Effect of Zero Delay

A delay value of zero is permitted in a LogicWorks device, but this setting should be used only with an understanding of how the simulation is implemented—as it can result in unexpected side effects.

Note that on a given pass through the simulation routine, all the events on the list which occur at the current time are scanned and then the new outputs for all affected devices are calculated. If any of these devices has a zero delay setting, then this will result in more changes being placed on the event list at the current time. However, all these changes emerging from zero–delay devices will not be evaluated until the next pass through the simulator. This is done to allow for user interaction with the simulation.

If you step interactively through a circuit with zero–delay elements, you will see all these value changes updated on the screen, even though "simulation time" does not advance. If a signal changes value and then reverts to its original state within the same time step, this will be displayed as a zero–width spike in the Timing window.

If a zero–delay feedback loop exists in a circuit, the signal changes will be simulated and any probes on the diagram will be updated at each pass through the simulator. However, the events at the head of the list will always have the same time value associated with them and the simulated time will never advance. This will stop the Timing window from updating until some delay is inserted in the loop.

Where Delays are Stored

For devices, the delay attribute field is called "Delay.Dev"; for pins, it is "Delay.Pin". An empty or invalid string will be interpreted as the default value, usually 1 for devices and 0 for pins.

Some special–purpose devices, such as the Clock and One Shot primitive devices, take two delay characteristics. In this case, two integers separated by a comma should appear in the Delay.Dev field.

◆ More information on this is given in the information section on each of these primitive types in Chapter 9, Primitive Devices.

Device Storage State

In LogicWorks, primitive storage devices (such as flip–flops, counters, and registers) do not store their current state internally. The device state is completely determined by the values on the signals attached to the output pins. Thus, the following factors will affect the operation of these devices:

▩ Conflicting or overriding values on the output signals (e.g., a stuck state) will override the last device state calculated by the model.

▩ Device and pin delays will influence the calculation of a new device state. For example, if the period of a clock applied to a counter is less than the total delay through it, an erroneous count sequence will result.

If desired, this behavior can be modified by placing the primitive devices in a subcircuit device and setting appropriate pin types and delays on the parent device to "buffer" the outputs.

NOTE: These comments do not apply to RAM or bidirectional switch primitives, both of which store internal state information independent of the values of the attached signals.

◆ See the section "Working With Subcircuit Devices" in Chapter 7, Simulation, for more information.

Input Signal Values

For all device types except switches, the signal values High Impedance and Conflict are treated as Don't Know when applied to a device input. When a device is first created, all input signals take the High Impedance state, and outputs are set depending on their type—normally to the Don't Know state. Thus an unused input pin will appear as an unknown input to a device, which may affect its output level.

As with real circuits, all unused inputs should be connected to a high or low level as appropriate. This can be done by naming the pin signal either "0" or "1", by using a power or ground symbol, or by using a pullup resistor to set a high level. See more information on logic states in other parts of this chapter.

Device Pin Types

Every device pin has a characteristic known as its *pin type*—for example, input or output. The pin type is set when the part entry in the library is created, and cannot be changed for individual device pins on the schematic. Correct pin type settings are crucial to correct and efficient operation of the simulator.

The pin type is used by the simulator to determine the direction of signal flow and to set the output values that are allowable on a given output pin.

◆ For detailed information on the available pin types and how they affect the simulation see Appendix B, Device Pin Types. For procedures for setting pin types when creating a symbol see Chapter 11, Device Symbol Editing.

Device Pin Inversion

The logic of any pin on any device can be inverted by placing a non–empty value in the Invert.Pin attribute field of the pin. When this is done, any value passing into or out from that pin will be inverted. This applies to primitive types as well as subcircuit devices. The following table summarizes the level mappings that occur.

External Signal Value	Internal Signal Value
LOW.H	HIGH.H
LOW.L	HIGH.L
HIGH.H	LOW.H
HIGH.L	LOW.L
All others	Unchanged

NOTE: 1) The logical inversion of the pin is *completely independent* of the graphical representation of the pin. For example, using the "inverted pin" graphic in the DevEditor *does not* invert the pin logic in the simulator. You must set the Invert.Pin field to have this effect.

2) Although pin inversion can be specified independently for each device on the schematic, we do not recommend modifying these settings after a device has been placed on the diagram. This can create the confusing situation of two devices with the same name and symbol but different logical characteristics.

See also:

◆ "Pin Delays and Inversion" on page 170, for information on pin inversion in subcircuit blocks.

◆ Chapter 9, Primitive Devices, for information on how pin inversion can be used with specific primitive types.

◆ Chapter 11, Device Symbol Editing, for procedures for setting pin attributes when creating a symbol.

Simulation Clearing and Initialization

The LogicWorks simulator provides a number of mechanisms to assist in setting initial values and resetting a simulation.

The Clear Simulation Operation

Windows—You can invoke the Clear Simulation operation by clicking on the Reset button ( ) in the Simulator toolbar.

Macintosh—You can invoke the Clear Simulation operation either by clicking on the Restart button on the Simulator Palette, or by selecting the Clear Simulation command in the Simulate menu.

This operation performs the following steps:

◆ Other tools (such as Timing) are notified and perform their own processing.

◆ All signal–change events on the queue are disposed of, whether pending or historical.

◆ Any clocks in the design are re-initialized.

◆ If any signal or pin initial values are specified, they are set up. See below for information on setting initial values.

◆ All devices are queued for immediate re-evaluation.

The Clear Unknowns Operation

The Clear Unknowns operation is a heuristic procedure which attempts to remove Don't Know signal values from a design. This can be used to find an initial state when a design is first simulated, or after any edit operations that result in unknown values.

Windows—You can invoke this operation by clicking on the Clear Unknowns ( ) button in the Simulator toolbar.

Macintosh—You can invoke this operation either by clicking on the Clear Unknowns button on the Simulator Palette, or by selecting the Clear Unknowns command in the Simulate menu.

The Clear Unknowns operation performs the following steps, stopping as soon as all unknown states are removed from the design:

◈ Any pending signal change that would result in an unknown state is removed from the queue.

◈ Any primitive type with storage capability (such as flip–flop, register, or counter) that has a Don't Know output value is cleared, either to its specified initial value (if any) or to zero.

◈ A single device that currently has an unknown output state is randomly selected and queued for re-evaluation. A special input mapping is done so that all unknown inputs are treated as zero.

◈ The simulator is cycled repeatedly as long as the number of unknown states in the design decreases.

◈ The last three steps are then repeated until the number of unknowns ceases to diminish.

If this operation does not clear the design to an appropriate state, refer to the other techniques discussed in following sections.

NOTE: Designs with "hard" unknowns, such as unconnected inputs or conflicting outputs, will not be successfully cleared by this procedure. All device inputs should be specified to a known value if not driven by other devices.

Setting Initial Values

You can specify initial values for signals and pins. These values will be applied by the Clear Simulation and Clear Unknowns operations, as described in the preceding sections.

For both object types, the initial value is entered into an attribute field, either Initial.Sig or Initial.Pin. The allowable values consist of a single character chosen from the following table.

Character	Value
0	LOW
1	HIGH
Z	HIGHZ
X	DONT01

All other values will be ignored.

NOTE: 1) It is left completely to the user to decide if the specified initial values make sense. No checking is done to determine if a given device output value is the reasonable result of the device's current input.

NOTE: 2) Devices do not have initial value settings, since their values are completely determined by the state of their output pins. See the section, Pin Initial Values, below.

Signal Initial Values

An initial value for a signal can be placed in the Initial.Sig attribute field using the format described in the previous section. When a Clear Simulation operation is invoked, the initial value specified is placed on the signal without regard for the current output levels of devices driving the signal. The given value will stay on the signal until some device driving the signal changes state, or some other user action changes it.

NOTE: If a pin initial value is specified for any output pin driving the signal, the signal value will be overridden.

Pin Initial Values

The initial value for a pin is stored in the Initial.Pin attribute field, using the format described earlier. Initial values can only be specified for output or bidirectional pins and will be ignored on input pins.

When a Clear Simulation operation is invoked, the specified initial value is placed on the pin without regard for the current inputs affecting the device. The given value will stay on the pin until the device model schedules a state change or some other user action changes it.

Schematic Simulation Issues

Working With Subcircuit Devices

The simulator does not impose any new rules on working with subcircuit devices, but editing a design with active simulation has some effects that should be noted.

◆ See also Chapter 6, Advanced Schematic Editing

Editing an Open Internal Circuit

A number of issues arise if you have used the same subcircuit device type multiple times in a design and you open one copy for editing (i.e., by using the Push Into command or by double–clicking on the device). You should note the following points:

▓ The Schematic tool creates a separate, temporary type definition for the open device when it is opened. Any simulation values that you view or change, or any circuit changes that you make, will *apply only to that one device* instance while it remains open.

▓ When you close an open internal circuit, the action taken depends on edits that have taken place. If you have made any edits (such as any graphical or structural change to the circuit) then all instance data (such as signal values, and so on) from other devices of the same type will be *lost*. It will be completely replaced by the values from the edited block.

The Port Interface

The connection between a pin on a parent device symbol and the corresponding signal in the internal circuit is quite complex, from a simulation standpoint. In order for this connection to act like a "hard wire" between the two levels, the following conditions must be met:

- The *pin type* on the parent device symbol must be "bidirectional."

- The *pin type* of the corresponding port connector in the internal circuit must be "bidirectional."

- The *pin delays* on *both* the pin on the parent device *and* the pin on the port connector must be zero.

- No *pin inversion* must be specified, either on the parent device pin or the port connector pin.

Any other combination of settings will result in some degree of isolation or "buffering" between the two levels. For example, The observed signal value on the signal in the internal circuit may be different from that on the parent pin.

NOTE: When a symbol is created in the DevEditor tool, all pins default to type "input"—that is, they will not drive any attached signal. If you are creating a subcircuit device symbol for simulation purposes, the pin types must be set to appropriate values.

The effects of these various settings are summarized in the following sections.

Parent Device Pin Type

Any signal value driven out of a parent pin by an internal circuit may be translated according to the pin type on the parent device. These effects are summarized in the following table.

Pin Type	Effect
Input	This will prevent that pin from ever driving the attached signal, regardless of drives in the internal circuit.
Output / Three–state	This will pass the sum of the internal drives up to the parent pin without any translation. Signal value changes on the signal attached to the parent pin *will not* be passed to the internal circuit.
Open collector / Open emitter	Any drive level from the internal circuit will be translated according the capability of the pin type. See Appendix B, Device Pin Types, for more details.
Bidirectional	All changes on the internal signal are passed to the parent pin and vice versa.
Other types	Other types, such as Tied High and Tied Low, are not recommended.

NOTE: Although it may be tempting to set all pins to "bidirectional," this is not recommended. It significantly increases simulation overhead and increases the difficulty of isolating circuit drive problems.

Port Connector Pin Type

The pin type on the port connector is also used to translate the value of any incoming signal changes, in a manner similar to the parent pin type. Normally, the pin type setting on a port connector should complement the setting of the parent pin, as follows:

Parent Pin Type	Port Connector Name	Port Connector Pin Type
Input	Port In	Output
Bidirectional	Port Bidir	Bidirectional
All others	Port Out	Input

Other settings on the port connector pin are not recommended.

Pin Delays and Inversion

The normal pin delay and inversion settings can be applied to the port interface. A non–null value in the Invert.Pin attribute field will cause any signal values passing in either direction to be inverted. An integer value in the Delay.Pin attribute will cause the specified delay to be inserted inline with level changes passing in either direction.

NOTE: 1) We recommend that pin delay and inversion settings be applied *only* to the pin on the parent device, and *not* to the port connector in the internal circuit. Attribute settings on the port connector are more difficult to verify and edit, since the port connector is a "pseudo–device" and some schematic editing operations will be disabled.

2) Changes made in the Invert.Pin and Delay.Pin attributes, after a device has been placed on the schematic, will affect only that one device instance. Default values can be set in these attribute fields when the symbol is created in the DevEditor.

Power and Ground Connectors

Power and Ground connector symbols do not have any inherent simulation signal drive, unless their pin type has been set to Tied High or Tied Low, as appropriate. The positive–supply symbols provided with LogicWorks have Tied High settings, while others will be Tied Low. The symbols provided with older LogicWorks releases may not have any drive setting, resulting in a high impedance level on these signals. This can be remedied by either:

- Replacing *any one* or *all* of the ground or power symbols with symbols containing the appropriate setting; or
- Forcing a Stuck High or Stuck Low level onto the signal, using the signal probe tool or the Stick Signals command. Note that, because all like–named ground or power segments are logically connected, this only needs to be done on a single segment.

Special Signal Names 0 and 1

The signal names 0 and 1 are recognized by the simulator as special. If any signal is named 0, it will be given a Stuck Low value. If a signal named 1 is found, it will be given a Stuck High value. These values can be cleared or changed using the signal probe, if desired.

◆ See the Signal Probe command in Chapter 12, Menu Reference, for more information.

Simulation Models

In order for LogicWorks to completely simulate a design, every symbol on the design must have an associated simulation model. In LogicWorks, simulation models can take one of the following forms:

■ **Primitive Devices:** These types have "hard–wired" program code to evaluate input and output changes. They include the gates, flip–flops, and other devices described in Chapter 9, Primitive Devices, as well as the user–definable PROM and PLA primitives.

■ **Subcircuit Devices:** The simulation function of a subcircuit device is completely determined by its internal circuit (except for the addition of pin delays and inversion). The definition of a device subcircuit can be stored with the part in a library. The subcircuit itself can contain any combination of primitive devices or other subcircuits (except itself, of course!) nested to any desired depth.

Whenever any device type is to be simulated, all information about the device must be loaded into memory. Unless you explicitly purge internal circuits or code models from the design, they will become permanent parts of the design and will be saved with the file.

Primitive Devices on the Schematic

The primitive devices provided in the primlogi.clf and primgate.clf libraries can be used at any time as part of a schematic, whether or not the simulator is installed. However, these libraries are not intended to match any real logic families and do not have any part name or pin number information associated with them.

◆ See Chapter 9, Primitive Devices, for more information on creating and using primitive types.

Simulation Pseudo–Devices

The simulation pseudo–devices (for example, those in the primio.clf library) are handled specially by the Schematic tool. In general, you cannot modify the symbols, pin types, or other characteristics of these devices. In addition, they are treated differently from normal device symbols in the following ways:

▓ By default, these devices are flagged "omit from report," meaning that they will not appear in any netlist or bill of materials reports. This setting can be changed using the Schematic tool's Get Info command.

▓ These symbols will not be assigned names when placed on a schematic. Names can be manually assigned, if desired.

The Switch and Keyboard types respond to a normal mouse click by changing state, rather than being selected. To select one of these devices, hold the (SHIFT) key pressed while clicking on it.

8
The Timing and Simulator Tools

The Timing Window

The Timing window allows you to display timing waveforms in graphical form and updates continuously and automatically as the simulation progresses. Only one Timing window can be displayed and it displays information for the active design. If multiple sub-circuit levels are open, all displayed waveforms are shown in a single window.

Windows—

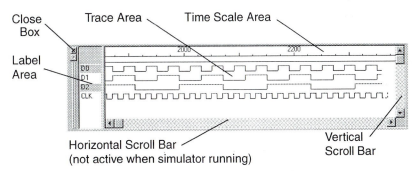

Close Box

Trace Area

Time Scale Area

Label Area

Horizontal Scroll Bar
(not active when simulator running)

Vertical Scroll Bar

Macintosh—

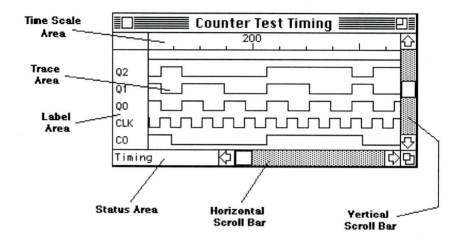

Here are the components of the Timing window:

Time Scale Area	Located just below the Timing window's title bar, the time scale is used to establish the absolute timing of value changes in the trace area. The scale is dependent upon the timing resolution (in Windows, set using the < and > buttons in the Simulator Palette; on the Macintosh, set using the <> and >< buttons in the Simulator Palette). The time scale is also used to set insertion points and selection intervals for use in editing functions.
Trace Area	This area displays simulation results and allows editing of waveforms. Waveforms can only be modified in the future, i.e., at times greater than the current simulation time.
Label Area	Displays the list of signal names corresponding to the timing traces at right. Traces can be repositioned by dragging them vertically in this area. In addition, a pop–up trace menu can be displayed in the Windows version by right–clicking in this area, and in the Macintosh version by ⌘–clicking in this area.
Status Area (Macintosh Only)	This box shows progress information for timing commands. Clicking in this box displays the Simulator Palette.
Horizontal Scroll Bar	This allows you to display time to the right or left of the present viewing area. The horizontal scroll bar is available when the simulation is stopped, but disabled when the simulation is running.

| **Vertical Scroll Bar** | This will display the signal labels and their corresponding traces above or below the ones presently displayed. |

Displaying Signals in the Timing Window

Adding a Signal Trace

To add one or more signal traces to the Timing window,

◆ Select any number of named signals in the schematic.

Windows—Click the Add to Timing tool (⊞) or select the Add to Timing command in the Simulation menu.

Macintosh—Select the Add to Timing command in the Simulation menu.

Removing a Signal Trace

To remove a trace from the Timing window,

◆ Select the traces to be removed by clicking in the label area of the timing window. You can remove multiple traces in one operation by holding the (SHIFT) key to select multiple labels.

Windows—Right–click on the selected name in the label area at the left side of the Timing window, then select the Remove command in the pop-up menu.

Macintosh—⌘–click on the name in the label area at the left side of the Timing window. Then select the Remove command in the pop-up menu.

Repositioning Traces

Any collection of selected labels and their corresponding timing traces can be repositioned within the list by clicking on the desired names—using the (SHIFT) key, if desired, to select more items—and dragging the outlined box vertically to its new location. Releasing the mouse button will cause the list to be revised with the labels and traces in their new positions. Alternatively, the To Top, To Bottom, and Collect commands in the Timing pop–up menu can be used.

Timing Display Groups

The Timing tool allows multiple signal lines to be grouped into a single trace with values displayed in hexadecimal.

Creating a Group Trace

A group trace can be created by either of these methods:

- Select any collection of traces by (SHIFT)–clicking in the label area, then select the Group command in the timing pop–up menu.

- Select a bus in the schematic diagram, then select either the Add to Timing command or Add as Group command in the Simulation menu. Busses are added as a group by default. They can then be ungrouped, if desired, using the Ungroup command in the timing pop–up menu.

Order Within a Group

For the purposes of displaying a hexadecimal value for a group, the order of signals within the group is important. When a group is created, the following rules are used to establish the order:

- If the signal name has a numeric part (e.g., D12 or WRDAT4X), then the numeric part is used to sort the signals. The lowest–numbered signal will be the least significant bit of the group value. Any unnumbered signals will be in the most significant bit positions.

- Otherwise, the signal's existing position is used—i.e., traces that appeared higher in the Timing window will be more significant.

The order of signals within a group can be changed using the Get Info command on a group trace. This is displayed by selecting the Get Info... command in the Timing pop–up menu or by double–clicking on the label.

Entering a Group Name

When a group is first created, a group name is automatically generated from the names of the enclosed signals. This name can be edited using the Get Info command in the Timing pop–up menu.

NOTE: The group name is lost when an Ungroup operation is performed.

The Windows Simulator Toolbar

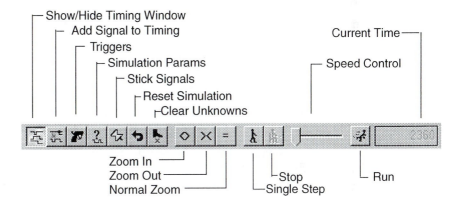

Displaying and Hiding the Simulator Toolbar

The Simulator toolbar is displayed by default when the Timing window is shown. You may move it or close it. To re-display the palette, select the Simulator Tools command from the View menu. To hide the toolbar, simply uncheck the same menu item.

NOTE: The Simulator toolbar can be displayed even if there is no Timing window displayed. This allows you to make use of the simulation controls even if you are not using the Timing window.

Simulator Toolbar Time Display

The status area of the Simulator toolbar displays one of two different time values, depending upon the status of the simulator:

▨ If the simulator is reset, it will display "0".

▨ Otherwise, it shows the current simulation time as the simulation progresses.

Simulator Toolbar Controls

The buttons in the Simulator toolbar control the simulator as follows:

Reset	Clears all pending events, sets time to zero and recalculates all device states.
Run	Causes the simulator to execute at the fastest possible speed.
Step	Causes the simulator to execute one time step.
< (Zoom In)	Increases horizontal display resolution in the Timing window, i.e., decreases number of time units per screen pixel.
= (Zoom Reset)	Resets zoom to the default level in the Timing window.
> (Zoom Out)	Decreases horizontal display resolution in the Timing window so more elapsed time can be viewed in the display
Trigger...	Displays the trigger control dialog.
Clear X	Clears all storage devices and attempts to clear feedback paths in the circuits.

The Macintosh Simulator Palette

Displaying and Hiding the Simulator Palette

The Simulator Palette is displayed by default when a new Timing window is created. It can be hidden by selecting the Hide Timing Palette command from the Simulate menu, or by clicking in its close box. To re-display the Palette, select the Show Timing Palette command from the Simulate menu, or click in the timing status area at the lower left corner of the Timing window.

NOTE: The Simulator Palette can be displayed even if there is no Timing window displayed. This allows you to make use of the simulation controls even if you are not using the Timing window.

Simulator Palette Time Display

The status box at the top left of the Simulator Palette displays one of two different time values, depending upon the status of the simulator:

- If the simulator is stopped and the cursor is positioned in the time scale or trace area of the Timing window, it shows the time corresponding to the cursor position.

- Otherwise, it shows the current simulation time as the simulation progresses. It cannot be disabled.

Simulator Palette Controls

The control buttons in the Simulator Palette provide quick access to functions that are also located in the Simulate menu.

	This speed control allows graphical selection of a simulator speed. Clicking in the gray area switches between Run and Stop. Clicking in the arrows or dragging the position box will select an intermediate speed.
Step	Causes the simulator to execute one time step.
Probe	This item is active only when the Schematic window is in front. It activates the Schematic signal probe mode.
Restart	Same as Clear Simulation menu command. Clears all pending events, sets time to zero, and recalculates all device states.
Clear Unknowns	Same as Clear Unknowns menu command. Clears all storage devices and attempts to clear feedback paths in circuit.

Trigger	Same as Trigger menu command. Displays the Trigger Setup dialog.
Disp Off/Disp On	Enables and disables the timing display. Simulation proceeds substantially faster when the timing display is disabled.
<> (Zoom In)	Increases horizontal display resolution, i.e., decreases number of time units per pixel.
>< (Zoom Out)	Decreases horizontal display resolution so more elapsed time can be viewed in the display.
Drawing Tools	The drawing tools are used for drawing and editing timing waveforms. These are described in the following section.

Trigger...

This command displays the Trigger Setup dialog, as illustrated below.

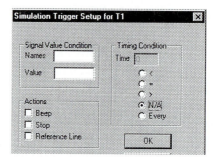

Trigger Conditions

The trigger is activated when two sets of conditions are met:

- The time condition—i.e., the current simulator time value —is less than, equal to, greater than, or a multiple of, a given value.
- Signal value condition, i.e., one or more signals are at specified levels.

Signal Value Condition Controls

The controls related to the signal condition are summarized in the following table:

Names   In this text box, you can type the names of one or more signals whose values will be compared to the hexadecimal integer value typed in the Value box. One or more signals can be entered using the following formats:

CLK   The single signal CLK
D7..0   The signals D7 (most significant bit), D6, D5...D0
IN1 OUT3   The signals IN1 and OUT3

Value   In this box, you enter the signal comparison value as a hexadecimal integer. This value is converted to binary and compared bit for bit with the signals named in the Names box. The rightmost signal name is compared with the least significant bit of the value, etc.

Time Condition Controls

The controls related to the time condition are summarized in the following table.

Time   In this text box, you enter the time value as a decimal integer. The meaning of this value is determined by the switches below it.

<, =, >   These buttons indicate that the trigger will be activated when the simulation time is less than, equal to, or greater than the given value, respectively.

N/A   This specifies that the time condition should be considered to be always true. The time value is ignored.

Every   This time option specifies that the trigger will be activated every time the simulator time equals a multiple of the specified value.

Trigger Actions

When the trigger is activated, any combination of the displayed actions can be invoked.

Beep   Generates a single system beep.

Stop   Stops the simulator immediately.

Reference Line Draws a reference line at this time on the Timing waveform
display.

Timing Window Editing

Macintosh—When the Timing window is active, the Macintosh Edit menu
provides the following commands for editing waveforms: Cut, Copy, Paste,
Clear, Duplicate, Select All, Insert Time, and Delete Time.

Windows—Right-clicking in the trace area of the Timing window will display a
pop-up menu. It provides the following commands: Copy, Paste, and Select
All (selects all traces).

NOTE: Timing traces can only be editing in the future—i.e., at times greater than
the current simulation time.

Selecting Data for Copy/Paste Operations

To select timing data for the editing operations described above:

▪ Simulation must be Stopped. To do this, use the speed control in the
Simulator Palette, select the Stop command in the Simulation Speed
submenu, or click anywhere in the Timing window.

▪ The cursor must be in Point mode. If not already in Point mode, click
on the arrow symbol in the Simulator Palette, or select the Point
command in the Edit menu. The cursor will now be an arrow.

There are two methods of selecting areas for edit operations.

Separate Label and Interval Selection

With this method, you select the traces to be affected by (SHIFT)–clicking in the label area; then select the time interval by clicking and dragging in the time scale. This allows you to select non-contiguous traces in the display.

◈ Click on the desired label in the label area to select it. To select more than one label, hold the (SHIFT) key and click on the labels.

◈ To set the selection interval, click and hold down the mouse button in the time scale at either end of the desired interval. Drag left or right until the desired interval is enclosed. When the mouse button is released, the select interval is set, and two selection interval lines will appear. If any of the signal labels were selected, the timing signal within the selected interval will be highlighted in the Timing window.

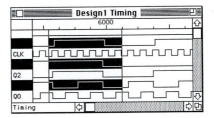

NOTE: Clicking and releasing the mouse button at one spot will create a zero–width interval. This can be used to insert Pasted data without deleting any existing data.

Drag Selection

This method allows you to select a group of labels and a time interval in a rectangular area of the Timing window.

To do a drag selection, click and hold the mouse button at any corner of the rectangular area you wish to select. Drag diagonally across the desired area. When the mouse button is released, the enclosed time interval and traces will be selected.

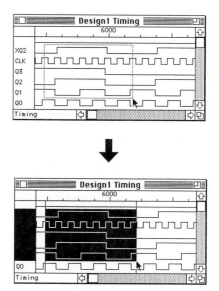

NOTE: The selection operations in the Timing window have no effect on selections in the Schematic window.

Selecting All Traces or All Time

To select a specific time interval in all traces on the diagram:

◆ Use the Select All command in the Edit menu to select the entire diagram.

◆ Drag–select an interval in the time scale area *without clicking in the trace area*.

To select all time for specific traces:

◆ Use the Select All command in the Edit menu to select the entire diagram.

◆ Click at the top of the label area, above the highest label displayed (this will deselect all traces). Then (SHIFT)–click to select the desired traces.

Deselecting

Clicking anywhere in the trace area that is *not* in a trace will deselect the labels and selection interval.

Clicking in the label area above or below the label list deselects all traces but leaves the current interval selected.

Summary of Timing Edit Commands

The following table summarizes operation of the Timing edit commands in the Edit menu.

Cut **(Macintosh only)**	This command copies to the Clipboard any signal–change events on selected signals in the selected time interval, and clears the selected interval. The data is stored in both picture form and text data form. Events after the selected interval *are not* moved forward. The Delete Time command can be used to do this.
Copy	The Copy command copies the selected timing data to the Clipboard in picture and text format. See the notes under the Cut command, above. Note that Copy *can* be used on a selection to the left of (older than) the current simulation time since it does not modify the selected data.
Paste	The Paste command pastes the text timing data from the Clipboard onto the selected area of the Timing window. The selected time interval is deleted and then the new data is inserted. That is, data following the selection interval will be moved forward by the width of the selection interval, then back by the width of the pasted data.
Clear **(Macintosh only)**	The Clear command clears any signal–change events in the selected area on the Timing window. Later events *are not* moved forward.
Duplicate **(Macintosh only)**	This command inserts a duplicate of all selected signal data on the Timing window after the selected interval. That is, the selected data is copied to a temporary location, then the selection point is moved to the end of the selected interval and the copied data is *inserted* at this point. All signal changes after the duplicate data are moved back in time by the width of the original selection.

Insert Time **(Macintosh only)**	This command inserts a blank time interval in the selected traces. The new interval is inserted in front of the selected interval and is of the same width as the selected interval.
Delete Time **(Macintosh only)**	This command deletes the selected time interval from the selected traces, and moves all later data ahead by the width of the interval.

◆ See Chapter 12, Menu Reference, for a detailed description of these commands. See Appendix D, Timing Text Data Format, for a description of the Clipboard data format.

NOTE: 1) If you wish to paste a timing picture into a word processing package, it may be necessary to first paste it into a drawing program to extract the picture data from the Clipboard. A word processing package will, by default, normally take the text data from the Clipboard.

2) You cannot modify timing data that is older than the current simulation time.

The Drawing Tools (Macintosh Only)

NOTE: The timing editing tools are available only in the Macintosh version of LogicWorks.

The Timing tools are used in editing the signal lines displayed in the Timing window. New states can only be created and modified in the future (to the right of the point where the simulation was halted).

+	**High/Low**	Lets you create the high or low signal state, depending on the vertical position of the cursor after you drag and release the mouse button.
		Lets you modify the time at which a change occurs by clicking on, and dragging horizontally, the vertical edge of the signal. Is set at the time where the mouse button is released.
		Lets you modify the state by clicking on a horizontal edge and dragging up or down. The new state is set when the mouse button is released.

▨	**Don't Know**	Lets you create the Don't Know signal state.
		Lets you modify the time at which the state occurs by clicking on, and dragging horizontally, the vertical edge of the signal. Is set at the time where the mouse button is released.
▨	**High Z**	Lets you create the high impedance signal state.
		Lets you modify the time at which the state occurs by clicking on, and dragging horizontally, the vertical edge of the signal. Is set at the time where the mouse button is released.
▨	**Conflict**	Lets you create the conflict signal state.
		Lets you modify the time at which the state occurs by clicking on, and dragging horizontally, the vertical edge of the signal. Is set at the time where the mouse button is released.
⊓	**Insertion**	Inserts only high/low signal states in between previously created states.

NOTE: 1) The simulator must be stopped to perform any timing editing operations.

2) The drawing tools have no effect on grouped lines. A group can be ungrouped into individual signals for editing—if desired—using the Ungroup command in the Timing pop–up menu.

9
Primitive Devices

Every device on a LogicWorks schematic has a characteristic known as its *primitive type*. The primitive type is set when the part entry in the library is created, and cannot be changed for individual devices on the schematic.

Primitive types fall into three general groups:

■ **Schematic symbols:** The two primitive types SUBCIRCUIT and SYMBOL fall into this category and are the normal primitive types used for creating schematic symbols. SUBCIRCUIT is the default type for symbols created using the DevEditor. There are no restrictions on the ordering or type of pins on these symbols.

■ **Pseudo–device types:** These are the symbols used for bus breakouts, power and ground symbols, etc.

IMPORTANT: LogicWorks has very specific requirements for the order and type of pins on pseudo–devices. Refer to Appendix A, Primitive Device Pin Summary, for information. These rules *are not checked* by the DevEditor.

■ **Simulation types:** The majority of the primitive types defined in the following tables are simulation primitives and are intended for use with the LogicWorks simulator.

IMPORTANT: The simulation primitive types should not be used for user–created symbols without a clear understanding of their function.

Schematic and Pseudo–Device Primitive Types

A small number of primitive types are used to distinguish the types of symbols used strictly for schematic diagramming purposes. These symbol types have no inherent simulation properties.

IMPORTANT: The pseudo–device types have specific pin order requirements that must be followed if you create one of these symbols using the DevEditor tool. Refer to Appendix A, Primitive Device Pin Summary, for more information.

Primitive Type	Description
SUBCIRCUIT	Symbol having an optional internal circuit. This is the default for symbols created using the DevEditor tool.
SYMBOL	Symbol with no internal circuit.
BREAKOUT	Splits signals out of or into a bus. These symbols are normally created using the New Breakout command in the Schematic menu, although they can be created using the DevEditor for special purposes.
SIGNAL CONNECTOR	Used for power and ground connections.
PORT CONNECTOR	Makes a connection between the signal to which it is connected and a like–named pin on the parent device.

Simulation Primitive Types

In LogicWorks primitive device types, the function of each pin is determined by its type (i.e., input or output) and by its sequential position in the device's Pin List (as seen when the part is opened in the DevEditor). Pin name is not significant. Each type has specific rules about the ordering of pins. Failure to adhere to these rules will result in incorrect simulator operation.

For many primitive types, certain control inputs and outputs can be omitted to create simplified device types. For example, on flip–flop types, the Set and Reset inputs can be omitted.

◈ See Appendix A, Primitive Device Pin Summary, for information on which combinations of inputs are allowable and on the required order.

The rest of this chapter provides information on these simulation primitive types. Because their simulation functions are hard–coded, they occupy much less memory space than subcircuit devices and simulate more efficiently.

NOTE: 1) In primitive devices, logic functions are associated with pins on a device symbol according to pin order. When creating primitive devices using the DevEditor tool, you must be aware of the pin order requirements for the device type you are using. Refer to the description of each type in this chapter and to Appendix A, Primitive Device Pin Summary.

2) Bus pins are *not supported* on primitive device types.

The following table lists the available primitives and their functions.

Primitive Type	Description	Related Type	Max. # Inputs
NOT	Inverter		1
AND	N–input AND gate	Any pin inversions	799
NAND	N–input NAND gate	Any pin inversions	799
OR	N–input OR gate	Any pin inversions	799
NOR	N–input NOR gate	Any pin inversions	799
XOR	N–input XOR gate	Any pin inversions	799
XNOR	N–input XNOR gate	Any pin inversions	799
Transmission Gate	Transmission Gate		1
Buffer	Non–inverting N–bit 3–state buffer with optional common inverted enable	Buffer	400
Resistor	Digital resistor		1
Multiplexer	M*N to M multiplexer		256
Decoder	1 to N line decoder		256
Adder	N–bit adder with carry in and out	Incrementer	256
Subtractor	N–bit subtractor with borrow in and out	Decrementer	256
D Flip–Flop	D–type flip–flop	optional S & R	1
D Flip–Flop with Enable	D–type flip–flop with clock enable	optional S & R	1
JK Flip–Flop	JK flip–flop	T flip–flop, optional S & R	1
Register	N–bit edge–triggered register		256
Counter	N–bit synchronous counter	Up/down	256

Primitive Type	Description	Related Type	Max. # Inputs
Shift Register	N–bit shift register		256
One Shot	Retriggerable one shot		1
Clock	Clock oscillator		1
Binary Switch	Debounced toggle switch		1
SPST Switch	Open/closed single pole switch		1
SPDT Switch	Double throw switch		1
Logic Probe	Signal level display		1
Hex Keyboard	Hexadecimal input device		1
Hex Display	Hexadecimal digit display		1
Unknown Detector	Unknown value detector		1

The following table lists devices supported primarily for compatibility with older versions of LogicWorks. We do not recommend using these in new designs.

Device	Description
Pullup	Pullup resistor, single pin
D Flip–Flop ni	D–type flip–flop (non–inv S & R)
JK Flip–Flop ni	JK–type flip–flop (non–inv S & R)
Glitch	Glitch detector (use Trigger mechanism now)
SimStop	Simulation halt device (use Trigger mechanism now)

Pin Inversion

In addition to the pin function options described in this chapter, any pin on any device can be inverted by specifying a value in the Invert.Pin attribute field. Any non–empty value will cause the pin logic to be inverted.

◆ See Chapter 7, Simulation, for more information

Gates and Buffers

The primgate.clf library contains the primitive gates that have a built–in simulation function. The NOT, AND, NAND, OR, NOR, XOR, and XNOR devices behave according to the appropriate truth tables for such devices. Any gate input which is in the Don't Know, High Impedance, or Conflict state is treated as a Don't Know. A gate with a Don't Know input will not necessarily produce a Don't Know output. For example, if one input of an AND gate is low, the output will be low, regardless of the state of the other input—as in the following truth table:

A	B	OUT
0	0	0
0	1	0
0	X	0
1	0	0
1	1	1
1	X	X
X	0	0
X	1	X
X	X	X

Gate Definition

The gate types, except NOT, can be created with any number of inputs from 0 to 799. They are defined as shown in the following table.

Function	Output is...	Output is DONT if...
AND	LOW if any input is low, HIGH otherwise	Some input is DONT and no input is LOW
NAND	HIGH if any input is LOW, LOW otherwise	Some input is DONT and no input is LOW
OR	HIGH if any input is HIGH, LOW otherwise	Some input is DONT and no input is HIGH
NOR	LOW if any input is HIGH, HIGH otherwise	Some input is DONT and no input is HIGH
XOR	HIGH if an odd number of HIGH inputs and no DONTs	Any input is DONT
XNOR	HIGH if an even number (or zero) of HIGH inputs and no DONTs	Any input is DONT

Gate Pin Order

The NOT type must have exactly one input and one output, in that order. All other logic gate types can have any number of inputs, up to the maximum LogicWorks limit of 800 pins, followed by a single output.

NOTE: Pin order is important in all primitive devices! When creating a gate type using the DevEditor tool, the output pin must be the *last item* on the pin list. See Appendix A, Primitive Device Pin Summary.

Pin Inversions

The logic of any pin on any device can be inverted by placing a non–empty value in the Invert.Pin attribute field of the pin.

For example, to create the following AND gate with one inverted pin:

...the following steps must be taken in the DevEditor tool:

◆ Create the desired graphic symbol using the DevEditor's drawing tools.

◆ Place the three pins as shown. *Order is important!* All primitive devices must have a specific pin order. For gates, all inputs come first and the output pin last.

◆ In the Pin Name List at left, double–click on the last pin (the output pin). This will display the Pin Information Palette for that pin.

◆ Set the pin type to Output. If desired, edit the pin name.

◆ Press the (ENTER) key to move to the next pin. You may use this technique to edit the other pin names (if desired) and to check that they are all set to Input.

◆ Close the Pin Information Palette.

◆ In the New Pin list, click once to select the input pin that is to be inverted. Then select the Pin Attributes command, which is located on the Options menu (Windows) or on the DevEdit menu (Macintosh).

◆ Select the Invert.Pin field in the Attributes Dialog.

◆ Enter the value "1" for this field, then click Done. (The actual value doesn't matter, as long as it is non–empty.)

◆ Select the Subcircuit / Part Type command on the Options menu (Windows) or the DevEdit menu (Macintosh).

◆ Click on the Set to Primitive Type button, then select the AND primitive type in the drop–down list.

◆ Close the PartType Configuration dialog and save the part to a library in the usual manner.

NOTE: 1) The logical inversion of the pin is *completely independent* of the graphical representation of the pin. That is, using the "inverted pin" graphic in the DevEditor *does not* invert the pin logic in the simulator. You must set the Invert.Pin field to invert the logic.

2) Inverted gate types NAND and NOR can be created by using the NAND and NOR primitive type settings. You can also use the AND and OR settings and either invert the output pin or invert the input pins (using DeMorgan's Theorem). These methods will produce identical simulation results. There is a slight memory overhead, but no execution–speed overhead, to using an inverted pin.

Transmission Gate

The transmission gate (X–Gate) device behaves as an electrically controlled SPST switch. When the control input is high, any level change occurring on one signal pin will be passed through to the other. Since the device has no drive capability of its own, it will behave differently than a typical logic device when a high impedance or low drive–level signal is applied to its signal inputs. Most other primitives, such as gates, interpret any applied input as either High, Low, or Don't Know. The transmission gate, on the other hand, will pass through exactly the drive level found on its opposite pin. Thus, a high impedance level on one pin will be transmitted as a high impedance level on the other pin. Note that the simulation of this device may produce unpredictable results in extreme cases, such as an unbroken ring of transmission gates.

NOTE: No variations in number or order of pins are possible with the XGATE primitive type. It must have exactly one control pin and two bidirectional pins, with pin order as described in Appendix A, Primitive Device Pin Summary.

Three–State Buffer

The three–state buffer has N data inputs, N data outputs, and an optional active–low enable input. If the enable input exists and is high, all outputs enter a High Impedance state. If the enable input doesn't exist or is low, each output will follow the corresponding input if it is low or high, or produce a Don't Know level otherwise.

NOTE: N is a placeholder. The limits on N differ depending on the device. See Appendix A for more detail.

A single–input three–state buffer is shown in the following table:

Enable	Data	Out
0	0	0
0	1	1
1	0	Z
1	1	Z

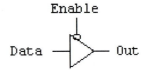

Making Non–Inverting Buffers

The Buffer primitive type can also be used to make a non–inverting buffer—that is, a buffer with its outputs always enabled—simply by omitting the enable input.

This can be used for the following purposes:

- To represent a non–inverting buffer or level translator in a design.
- To insert a delay in a signal path without affecting the logic of the signal.
- To create various types of open collector, open emitter, or inverting buffers, when used in conjunction with different pin type and inversion settings on the outputs.

NOTE: It is more efficient to use the NOT primitive type to make a simple inverter.

Resistor

The resistor device simulates the effects of a resistor in a digital circuit. It is more general than the Pullup Resistor device and can be used as a pullup, pulldown, or series resistor. Whenever a signal–level change occurs on either pin of the resistor, the device converts that level into a resistive drive level (see Chapter 7, Simulation, for more information on drive levels). A high impedance drive on one end is transmitted as a high impedance output to the other end. Note that LogicWorks does not simulate analog properties of devices, so the resistor device does not have a resistance value in the analog sense and will not interact with capacitor symbols placed on the same line. The effect of resistance on line delay can be simulated by setting the delay of the resistor device.

Logic Devices

Multiplexer

This is a device that selects one of N data inputs and routes it to a corresponding output line. There can from 1 to 256 outputs, plus an optional enable input, as long as the total pin count does not exceed the 800–pin limit.

A typical 8–to–1 multiplexer obeys the following function table, in which X = Don't Care:

EN	S2	S1	S0	D7	D6	D5	D4	D3	D2	D1	D0	Q0
0	0	0	0	X	X	X	X	X	X	X	0	0
0	0	0	0	X	X	X	X	X	X	X	1	1
0	0	0	1	X	X	X	X	X	X	0	X	0
0	0	0	1	X	X	X	X	X	X	1	X	1
0	0	1	0	X	X	X	X	X	0	X	X	0
0	0	1	0	X	X	X	X	X	1	X	X	1
0	0	1	1	X	X	X	X	0	X	X	X	0
0	0	1	1	X	X	X	X	1	X	X	X	1
0	1	0	0	X	X	X	0	X	X	X	X	0
0	1	0	0	X	X	X	1	X	X	X	X	1
0	1	0	1	X	X	0	X	X	X	X	X	0
0	1	0	1	X	X	1	X	X	X	X	X	1
0	1	1	0	X	0	X	X	X	X	X	X	0
0	1	1	0	X	1	X	X	X	X	X	X	1
0	1	1	1	0	X	X	X	X	X	X	X	0
0	1	1	1	1	X	X	X	X	X	X	X	1
1	X	X	X	X	X	X	X	X	X	X	X	1

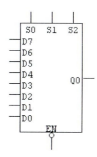

Multiplexer Pin Variations

A number of variations in multiplexer logic are possible with this primitive type, depending on which input and output pins are included. The following table summarizes the possible variations. Samples are shown with M=1 and N=2, but any combination of M and N can be used within the maximum pin limit of 800.

Number of Sections	Number of Inputs/Section	Number of Select Inputs	Number of Enable Inputs	Sample Symbol
M	2^N	N	0	
M	2^N	N	1	
M	$2^{N-1}+1 .. 2^N$	N	0*	

* If there are fewer than 2^N inputs per section, there can be no enable input.

❖ Specific pin order requirements for the multiplexer type are given in Appendix A, Primitive Device Pin Summary.

Decoder

The Decoder (active low) primitive device activates one of N outputs depending on M select inputs, as follows (X = Don't Care):

EN	S2	S1	S0	7	6	5	4	3	2	1	0
0	0	0	0	1	1	1	1	1	1	1	0
0	0	0	1	1	1	1	1	1	1	0	1
0	0	1	0	1	1	1	1	1	0	1	1
0	0	1	1	1	1	1	1	0	1	1	1
0	1	0	0	1	1	1	0	1	1	1	1
0	1	0	1	1	1	0	1	1	1	1	1
0	1	1	0	1	0	1	1	1	1	1	1
0	1	1	1	0	1	1	1	1	1	1	1
1	X	X	X	1	1	1	1	1	1	1	1

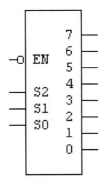

Adder/Incrementer

The N–bit Adder accepts one or two N–bit input arguments and (optionally) a 1–bit carry, and outputs their N–bit sum plus an optional 1–bit carry out.

Multiple Adders can be connected together by feeding the Carry Out from each stage to the Carry In of the next more significant stage. The Carry In to the least significant stage should be set to zero.

Adder Pin Variations

The adder primitive can be used in four variations, as summarized in the following table. Sample symbols are shown with 4–bit inputs, although any number of bits from 1 to 256 is permissible.

	Has B Inputs	**No B Inputs**
Has Carry In	$S = A + B + Cin$	$S = A + Cin$
No Carry In	$S = A + B$	$S = A + 1$

In addition, the Carry Out pin can be independently included in, or omitted from, any of these configurations.

◆ Refer to Appendix A, Primitive Device Pin Summary, for precise pin order requirements.

Subtractor/Decrementer

The Subtractor primitive type behaves identically to the Adder type except that a subtract or decrement operation is performed, depending upon pin configuration.

D Flip–Flop

The D–type flip–flop is positive-edge–triggered and obeys the following function table:

S	R	D	Clock	Q	Q/
0	0	X	X	1	1
0	1	X	X	1	0
1	0	X	X	0	1
1	1	0	Rises	0	1
1	1	1	Rises	1	0
Rises	Rises	X	X	X	X

In the above table, X on the input side means Don't Care and on the output side means Don't Know.

Flip–Flop Setup and Hold Times

None of the LogicWorks primitive types explicitly implement variable setup and hold times. However, all edge–triggered primitives have an effective setup time of 1 unit, since they always use the input signal value existing before the current step. For example, if the data input changes at the same time as the clock, the old data value will be used to determine the new output value.

You can modify this effective setup time by specifying input pin delays on either the data or clock pins. You can check for setup and hold violations by using the simulator's Trigger capability to watch for value changes within a set amount of time.

Flip–Flop Initialization

Note that when a flip–flop is first placed in the schematic, it is in an unknown state and must be correctly initialized before it will produce predictable outputs. This can be done in the following ways:

■ Adding circuitry to force an explicit reset.

■ Using the Clear Unknowns button or menu command to force an initial state before starting the simulation.

■ Specifying an initial output value for *both* the Q and Q/ outputs in their respective Initial.Pin attributes. This will be applied every time a Clear Simulation command is executed.

D Flip–Flop Optional Pins

The D Flip–Flop primitive type has the following optional pins:

■ The Q/ (Not–Q) output can always be omitted.

■ The Set input alone, or *both* the Set(S) and Clear(C) inputs, can be omitted.

◆ Refer to Appendix A, Primitive Device Pin Summary, for specific pin order information.

D Latch

The D Latch primitive type is identical to the D Flip–Flop in function and pin specifications, except that it is level–triggered instead of edge–triggered. For example, the Q and Q/ outputs will follow the level of the D input as long as R is high.

D Flip–Flop with Enable

The D–type flip–flop with Enable is identical to the D Flip–Flop in function, except that it has an added active–high clock enable input. This input must be high at the time of the rising edge on the clock input for the data at the D input to be passed to the Q output.

JK Flip–Flop

The JK flip–flop is negative-edge–triggered and obeys the following function table:

S	R	J	K	Clock	Old Q	New Q	New Q/
0	0	X	X	X	X	1	1
0	1	X	X	X	X	1	0
1	0	X	X	X	X	0	1
1	1	0	0	falls	0	0	1
1	1	0	0	falls	1	1	0
1	1	0	1	falls	X	0	1
1	1	1	0	falls	X	1	0
1	1	1	1	falls	0	1	0
1	1	1	1	falls	1	0	1
rises	rises	X	X	X	X	X	X

In the above table, X on the input side means Don't Care and on the output side means Don't Know.

If any inputs are in an unknown state, the simulator will determine the output state where possible, or else set it to Don't Know.

◆ See the notes under D Flip–Flop, above, on setup and hold times and initialization.

Register

This device implements an N–bit, positive-edge–triggered register, with common clock and optional active–high clear inputs.

◆ See the comments on Setup and Hold times and initialization in the D Flip–Flop section, earlier in this chapter.

The following table illustrates some pin variations available for the Register primitive type:

4–bit register with active–high clear

4–bit register with active–low clear (using pin inversion)

4–bit register without clear

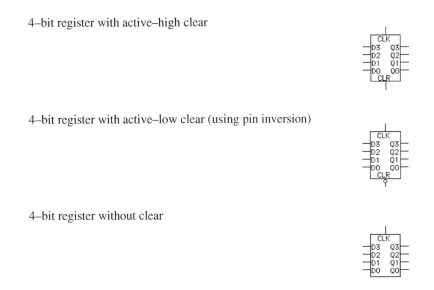

Counter

This device implements an N–bit, presettable, synchronous, positive-edge–triggered, up/down counter with active–low enable. The load data inputs and most of the control inputs can be omitted for simplified versions.

The following timing diagram shows a typical count cycle. Note that the CO (Carry Out) output goes low when the count reaches 2^N-1 (when counting up) or 0 (when counting down), and rises again on the next count. This can be used to cascade multiple counters together, as shown. The CLR input clears the counter asynchronously (that is, regardless of the state of the clock). The Count/Load input, when low, causes the data from the N data inputs (D0–D3) to be passed to the outputs (Q0–Q3) on the rising edge of the next clock. The Enable input disables counting when high, but has no effect on loading.

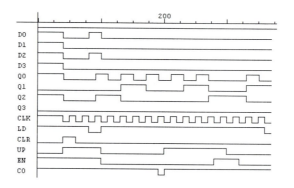

Cascading Multiple Counters

Counter primitives with the optional Enable and Carry Out pins can be cascaded to form larger synchronous counters as follows:

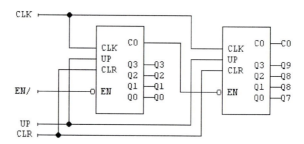

Counter Pin Variations

The following table summarizes the possible pin usage variations for the counter primitive type. The samples are shown with N=4, although the number of bits can be anywhere in the range 1 to 256.

Optional Inputs	**Including Load Inputs**	**Excluding Load Inputs**
CLR, UP/DN, ENABLE	CLK, UP, CO, CLR, D3, Q3, D2, Q2, D1, Q1, D0, Q0, LD, EN	CO, CLK, Q3, UP, Q2, CLR, Q1, EN, Q0
CLR, UP/DN	CLK, UP, CO, CLR, D3, Q3, D2, Q2, D1, Q1, D0, Q0, LD	CO, CLK, Q3, UP, Q2, CLR, Q1, Q0
CLR	CLK, CO, D3, Q3, D2, Q2, D1, Q1, D0, Q0, LD, CLR	CO, CLK, Q3, Q2, CLR, Q1, Q0
none	CLK, CO, D3, Q3, D2, Q2, D1, Q1, D0, Q0, LD	Q3, Q2, Q1, CLK, Q0

NOTE: CO can be independently included or omitted in any of the above variations.

Shift Register

The shift register is an N–bit, positive-edge–triggered device with serial or optional parallel load. When the Shift/Load input is low, data from the N parallel data input lines is transferred to the outputs on the rising edge of the next clock. When Shift/Load is high, the next rising clock edge causes the value at the Shift In input (SI) to become the new value for output Q0, as Q0 shifts to Q1, Q1 to Q2, etc., and the old value at the most significant output is lost.

The following table shows the shift register primitive with and without parallel inputs.

With Parallel Load   **Without Parallel Load**

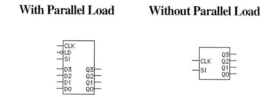

NOTE: The Shift Register primitive cannot be created without data outputs (that is, as a parallel–in, serial–out register) because the flip–flop values are stored on the output pins. Primitive devices have no internal state storage. See more comments on this in Chapter 7, Simulation.

Clock

The clock oscillator is used to generate a repeating signal to activate other devices. When it is first created, the clock output pin will be low; then after a delay time called the "low time," it will change to the high state. After a further delay called the "high time," the signal will revert to low and the cycle will repeat. The low and high times are initially set to 10, but can be modified:

Windows—Select the Parameters command from the Simulation menu.

Macintosh—Select the Set Params command from the Simulate menu.

Any number of Clocks may exist at once with independent delay times.

Creating Synchronized or Offset Clocks

When the Clear Simulation operation is selected (via the Reset button on the Simulator Palette), all clocks in the design are restarted. Clock outputs will be set to the low state and the timer for the low period will be restarted. Clock high and low times, combined with pin inversion and pin delay settings, can be used to precisely determine the relationship between two clock outputs. The following circuit example summarizes these options.

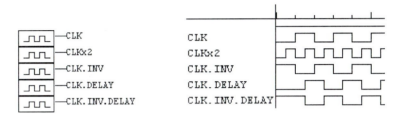

Signal	Low Time	High Time	Invert.Pin	Pin Delay
CLK	10	10		0
CLKx2	5	5		0
CLK.INV	10	10	1	0
CLK.DELAY	10	10		5
CLK.INV.DELAY	10	10	1	5

Setting Clock Values

To set the high and low times for a clock, first select the device in question (by activating the arrow cursor and clicking inside the device symbol), then choosing the Windows Parameters item on the Simulator menu or the Macintosh Simulation Params item on the Simulate menu.

You will be presented with a dialog box with buttons for increasing or decreasing the high and low values. The minimum for either value is 1 and the maximum is 32,767.

◆ See more information on the Parameters command in Chapter 12, Menu Reference.

One Shot

The One Shot is used to generate an output pulse of a fixed length when it is triggered by the rising edge of the trigger input. Two parameters can be set for a One Shot: the delay from the rising edge of the input to the start of the output pulse, and the duration of the pulse. The delay and duration times are initially set to 1 and 10, respectively, but can be modified using the Windows Parameters item on the Simulator menu or the Macintosh Simulation Params item on the Simulate menu.

The One Shot device is retriggerable, meaning that the output pulse will not end until *duration* time units have passed since the last trigger input. Repeating the trigger input can cause the output pulse to be extended indefinitely.

Setting One Shot Values

To set the delay and duration times for a One Shot, first select the device in question, then choose the Windows Parameters command on the Simulator menu or the Macintosh Simulation Params command on the Simulate menu.

◆   Refer to Chapter 12, Menu Reference, for more information on the Parameters command.

I/O Simulation Pseudo–Devices

Binary Switch

The Binary Switch device provides a means for setting a signal to a low or high level. When a switch is first created, its output is at a low level. Activating the arrow cursor and clicking on the switch causes the switch arm to move and the output to change to the opposite state. Any number of device inputs can be driven by a switch output. A switch has no delay characteristic since it has no inputs.

To select a switch device, rather than change its state, hold down the (SHIFT) key on the keyboard while clicking on the device.

SPST Switch

The SPST switch device simulates the actions of a simple open/closed switch in a digital circuit. When a switch is first created, it is open, and both connections present a high impedance logic level. Clicking on the switch (between the two dots) with the cursor in Point mode causes the switch arm to close and the switch to "conduct." In terms of the digital simulation, this means that whatever logic level is present on each pin is transmitted to the other one.

An SPST switch has a default delay of zero but this can be set to any value from 0 to 32,767 using the Windows Parameters command or the Macintosh Simulation Params command.

SPDT Switch

The SPDT switch device operates in essentially the same manner as the SPST switch described above, except that it always conducts between the single pin on one side and one of the two pins on the other. As with the other two switch types, clicking on it with the arrow cursor changes the position of the contact.

SPDT Pushbutton

The Windows SPDT Pushbutton switch device operates in essentially the same manner as the SPDT switch described above, except that it only stays switched to one side while the mouse is pressed.

Binary Probe

The Binary Probe is a device for displaying the level present on any signal line. When the probe is first created, its input is unconnected and therefore in the High Impedance state, which will be displayed as a "Z". When the input pin is connected to another signal, the displayed character will change to reflect the new signal's current state. Any further changes in the signal state will be shown on the probe. Possible displayed values are 0 (low) and 1 (high). In the Macintosh version, additional possible displayed values include X (Don't Know), Z (High Impedance), or C (Conflict).

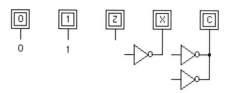

Hex Keyboard

The hex keyboard outputs the binary equivalent of a hexadecimal digit on four binary lines. A "key" is pressed by positioning the tip of the arrow cursor in the desired key number and clicking the mouse button. The binary data on the output lines will change to reflect the new value and will remain set until the next key is pressed. The fifth output line will go high momentarily and then low again when a key is pressed.

Hex Display

The hex display shows the hexadecimal equivalent of its four binary inputs. If any of the inputs is unknown, high impedance, or conflict, an X will be displayed.

◈   See Chapter 7, Simulation, for more information on pin delay and
    inversion.

10
RAMs and Programmable Devices

This chapter provides details on creating and using RAM (Random Access Memory), PROM (Programmable Read-Only Memory) and PLA (Programmable Logic Array) devices with user-specified data.

These devices are created using the PROM/RAM/PLA Wizard on Windows systems or the PROM/RAM/PLA Assistant on Macintosh systems. Apart from this slight difference in terminology, procedures are essentially the same on both systems. These terms will be used interchangeably in the rest of this chapter.

The RAM, PROM and PLA Primitive Types

LogicWorks supports the direct simulation of RAM, PROM and PLA devices as primitives. This means that you can efficiently represent each of these devices as a single simulation device, rather than having to generate a circuit built of equivalent logic devices.

The RAM, PROM and PLA devices represent "raw" memory or PLA (AND-OR) arrays. The primitive device models do not include capability for registers, feedback, three-state buffers, or other device features. In order to model these features in industry-standard PLD and PROM types, the PLD tool automatically generates a subcircuit model made up of a raw PLA device plus other primitive registers and buffers (etc.) as needed. The input to the PLD tool is a file that describes the structure of the device. The format of this file is described in on-line documentation provided with LogicWorks.

RAM Device Characteristics

The RAM primitive device supports the direct simulation of static Random Access Memory devices in a variety of configurations. You can create custom RAM devices with a variety of pin options.

This table summarizes the options available in the RAM primitive type.

Chip Enables	0, 1, 2, or 3 active–low chip enables. If any chip–enable input is high, all read and write functions are disabled. If no enable is provided, the device will always be enabled.
Write Enable	The active–low Write Enable pin is not optional. A low level on the pin causes the data present at the Data In lines to be written to the location selected by the address lines.
Output Enable Pin	This active–low pin controls output enable but does not affect writing.
Data In/Out	The data input and output lines can either be separate or can be combined into a single I/O bus.
Three–State Outputs	If the input and output lines are combined, or if three–state outputs are specified, the outputs enter a high impedance state if Write Enable or any Chip Enable is high. If three–state outputs are not specified, data outputs will be high when disabled.
Single–Word Simulation	If this option is selected, only a single word of real memory will be allocated for simulation purposes (i.e., the address inputs will be ignored). This allows logic testing of a circuit containing a large RAM device without consuming large amounts of program memory.
Common I/O	This option specifies that a single I/O pin will be used per data bit, rather than separate data in and data out lines. In this case, three–state outputs are assumed and outputs will be disabled when writing.

Don't Know Input Handling in RAM Devices

If any combination of Don't Know values on the control inputs could cause a write, then the selected memory location will be invalidated (that is, the location will contain Don't Know values). If the address inputs also have Don't Know values, then the entire device will be invalidated.

RAM Pin Delay and Inversion Options

The normal options for pin delay (using the Delay.Pin attribute field) and pin inversion (using the Invert.Pin attribute field) can be used with RAM devices.

RAM Device Limitations

RAM devices must fall within all of the following limits:

- 30 address-line inputs.
- 256 bits per word.
- Total memory space $< 2^{31}$ bytes.
- Sufficient program memory free to allocate a block twice the size of the simulated memory space.

NOTE: The Single Word Simulation option allows you to simulate a device with a large number of address inputs without having to allocate memory for all possible memory locations.

PROM Device Characteristics

For the purposes of simulation in LogicWorks, a PROM (Programmable Read Only Memory) is defined as a device having N inputs (from 1 to 30) and M outputs (from 1 to 256), and having 2^N storage locations, each containing M bits. Each different input combination selects one of the storage locations, the contents of which appear on the output lines. The number of storage locations required doubles for each input bit added, so PROM organization is only practical for a relatively small number of inputs. The advantage of the PROM is that any arbitrary Boolean function can be represented simply by storing the truth table for the function in the appropriate storage locations.

PROM Size Limits

PROM devices must fall within all of the following limits:

- 30 address-line inputs.
- 256 bits per word.

■ Total PROM memory space $< 2^{31}$ bytes.

■ Sufficient program memory free to allocate a block twice the size of the simulated memory space.

PLA Device Characteristics

In LogicWorks, a PLA (Programmable Logic Array) models a group of AND gates feeding into a single OR (active high) or NOR (active low) gate for each output bit. Each AND-gate input is connected to either an input bit, the inverse of an input bit, or constant high. By selectively making these input connections, it is possible to determine which input combinations will produce 0s or 1s in the outputs. PLAs are actually represented internally in a compact binary format, not as a netlist of AND and OR gates.

The input connections required to implement simple logic functions can generally be determined "by eye" for simple cases, whereas more complex logic must be reduced using Karnaugh maps, the Quine-McClusky method, or other more advanced design techniques. These methods are discussed in numerous circuit design textbooks and will not be covered here. Logic-Works has the capability of reading device data produced by external logic compiler programs.

PLA Size Limits

PLA devices must fall within the following limits:

■ Number of Inputs: 1 to 128

■ Number of Outputs: 1 to 128.

■ Number of product terms per output <= 65,535.

Complex Programmable Logic Devices

The term Programmable Logic Device (as opposed to Programmable Logic *Array*) will be used here to refer to a real programmable device that consists of one or more AND-OR planes plus associated registers, buffers, feedback paths, and so on. There is no PLD "primitive" device in Logic-Works. Some very simple PLDs can be directly simulated with a single

PLA primitive type—for example, a PAL10L8-type device. However, most PLDs are simulated by creating a subcircuit containing one or more PLA primitive devices, plus the other required logic and wiring. In most cases, these subcircuits are created automatically by the PLD tool, which is described later in this chapter, or by manually creating a subcircuit model of the target device using a PLA primitive device for the core.

Using the PROM/RAM/PLA Wizard

The PROM/RAM/PLA Wizard guides you through the steps for defining one of these complex primitive types and provides several alternate methods of entering data.

Creating a RAM Device

To create a RAM device, follow these steps:

Windows—Open the PROM/RAM/PLA Wizard by clicking on its button (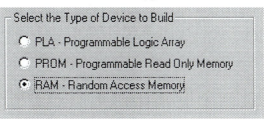) in the toolbar.

Macintosh—Open the PROM/RAM/PLA Assistant by selecting the PROM_PLA_RAM item in the Tools menu.

◆ Select the RAM device type and click the Next button:

Select the Type of Device to Build

- ○ PLA - Programmable Logic Array
- ○ PROM - Programmable Read Only Memory
- ● RAM - Random Access Memory

◆ Enter the desired number of inputs and outputs and select the appropriate options. For more information on these options, see "RAM Device Characteristics" on page 216. Click the Next button when done.

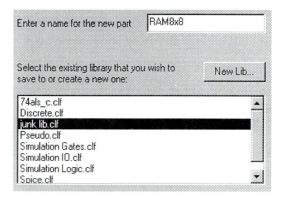

◆ Enter the name that you wish the new part to be saved under and select the destination library. If you need to create a new library, click the New Lib button.

Enter a name for the new part   RAM8x8

Select the existing library that you wish to   New Lib...
save to or create a new one:

74als_c.clf
Discrete.clf
junk lib.clf
Pseudo.clf
Simulation Gates.clf
Simulation IO.clf
Simulation Logic.clf
Spice.clf

NOTE: We do not recommend saving your own parts to any of the standard LogicWorks release libraries. This greatly complicates upgrading to new versions of the package.

◆ Click the Finish button to save the completed part.

The part is now ready to use by double-clicking on the new item in the parts palette. You can also make any desired graphical or attribute changes to the new part using the device symbol editor.

Creating a PROM Device from a Data File

This section describes the steps necessary to create a PROM device and read its contents from a data file. Two file formats are supported:

■ Intel hex format—This is a structured format generated by many assemblers and development systems. This format is more reliable because it includes a checksum, but is not practical to generate by hand.

■ Raw hex format—This is a free format that allows small devices to be easily defined manually or with simple software tools.

These formats are described in the on-line documentation provided with LogicWorks and in the Format Help accessible through the PROM/RAM/PLA Wizard.

Windows—Open the PROM/RAM/PLA Wizard by clicking on its button (⊞) in the toolbar.

Macintosh—Open the PROM/RAM/PLA Assistant by selecting the PROM_PLA_RAM item in the Tools menu.

◆ Select the PROM device type and click the Next button:

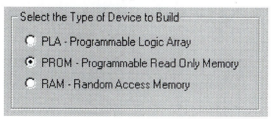

◈ Enter the desired number of inputs and outputs and select the "Intel-format hex" or "raw hex" data-entry method, as appropriate. Click the Next button.

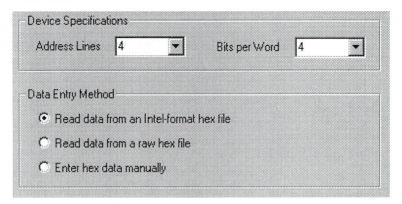

◈ Click the "Select Intel Hex File" or "Select Raw Hex File" button and locate the desired input file. The Format Help button can be used to bring up a description of the selected format.

◈ Click the "Next" button.

◈ Enter the name that you wish the new part to be saved under and select the destination library. If you need to create a new library for the part, click the New Lib button.

NOTE: We do not recommend saving your own parts to any of the standard LogicWorks release libraries. This greatly complicates upgrading to new versions of the package.

◈ Click the Finish button to save the completed part.

The part is now ready to use by double-clicking on the new item in the parts palette. You can also make any desired graphical or attribute changes to the new part using the device symbol editor.

Creating a PROM Device with Manual Data Entry

This section describes the steps necessary to create a PROM device and directly enter the data that will be stored in it. Here is a summary of the format used to enter PROM hex data:

- Each string of hexadecimal characters (0–9, a–f, A–F) specifies one word in the array. Words are entered starting with address 0, then address 1, and so on.

- Any non-hexadecimal character (including blanks, non-hex letters, punctuation and line breaks) separates one word from the next.

- Each hex character represents 4 bits, with the rightmost character representing the least significant bits in the word (i.e., bits 3, 2, 1, and 0).

- If insufficient hex characters are given for the length of the word, the rightmost character in the group represents the least significant bits and unspecified higher order bits are filled with zeros.

- If all the words in the device are not specified, unspecified words are filled with Don't Know (X).

- Line breaks can be inserted wherever desired except in the middle of a word.

- Items not containing any hex characters will be completely ignored and just taken as separators. Don't consider this to be a method of inserting comments unless you're sure you can spell everything without A to F!

- There is no comment mechanism.

- No error messages are given no matter what you put in the text!

To create the PROM device:

Windows—Open the PROM/RAM/PLA Wizard by clicking on its button ( ) in the toolbar.

Macintosh—Open the PROM/RAM/PLA Assistant by selecting the PROM_PLA_RAM item in the Tools menu.

◆ Select the PROM device type and click the Next button:

Select the Type of Device to Build

○ PLA - Programmable Logic Array

◉ PROM - Programmable Read Only Memory

○ RAM - Random Access Memory

◆ Enter the desired number of inputs and outputs and select the "Enter hex data manually" data entry method. Click the Next button.

◆ Enter the hex data in the text box provided, using the format described earlier. The Format Help button can be used to bring up a description of the format along with some examples.

◆ Click the "Next" button.

◆ Enter the name that you wish the new part to be saved under and select the destination library. If you need to create a new library, click the New Lib button.

NOTE: We do not recommend saving your own parts to any of the standard LogicWorks release libraries. This greatly complicates upgrading to new versions of the package.

◆ Click the Finish button to save the completed part.

The part is now ready to use by double-clicking on the new item in the parts palette. You can also make any desired graphical or attribute changes to the new part using the device symbol editor.

Creating a PLA from a Data File

This section describes the steps necessary to create a PLA device and read its contents from a data file. The file format used is called a DWL (Design-Works Link) file, and its format is described in on-line documentation provided with LogicWorks and in the Format Help accessible through the PROM/RAM/PLA Wizard.

Windows—Open the PROM/RAM/PLA Wizard by clicking on its button (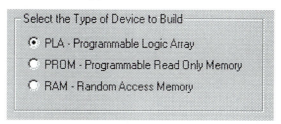) in the toolbar.

Macintosh—Open the PROM/RAM/PLA Assistant by selecting the PROM_PLA_RAM item in the Tools menu.

◆ Select the PLA device type and click the Next button:

Select the Type of Device to Build
- ◉ PLA - Programmable Logic Array
- ○ PROM - Programmable Read Only Memory
- ○ RAM - Random Access Memory

◆ Enter the desired number of inputs and outputs and select the "Read data from a DWL-format PLA file" data-entry method. Click the Next button.

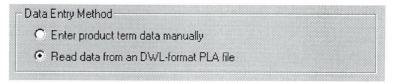
Data Entry Method
- ○ Enter product term data manually
- ◉ Read data from an DWL-format PLA file

◆ Click the "Select DWL File" button and locate the desired input file. The Format Help button can be used to bring up a description of the DWL format.

◆ Click the "Next" button.

◆ Enter the name that you wish the new part to be saved under and select the destination library. If you need to create a new library, click the New Lib button.

NOTE: We do not recommend saving your own parts to any of the standard LogicWorks release libraries. This greatly complicates upgrading to new versions of the package.

◆ Click the Finish button to save the completed part.

The part is now ready to use by double-clicking on the new item in the parts palette. You can also make any desired graphical or attribute changes to the new part using the device symbol editor.

Creating a PLA Device with Manual Data Entry

This section describes the steps necessary to create a PLA device and directly enter the data that will be stored in it: Here is a summary of the format used to enter PLA data:

■ Each PLA output is specified separately. The Wizard will step you through the outputs one page at a time, starting with the least significant output bit.

■ Each line of text entered represents one product term (AND function). Any number of product terms can be entered, including zero. If any one of the product terms specified matches the input values, the output will become active.

■ Each product term line must consist of N characters, where N is the number of device inputs. Whitespace characters are ignored and are not included in this count. The first character on a line corresponds to the most significant device input, the last character to the least significant.

■ Each input character must be one of the following

    0 Active if corresponding input is low

    1 Active if corresponding input is high

  X or x Always active

■ There is no comment mechanism.

■ No error messages are given no matter what you put in the text!

To create the PLA device:

Windows—Open the PROM/RAM/PLA Wizard by clicking on its button (⊞) in the toolbar.

Macintosh—Open the PROM/RAM/PLA Assistant by selecting the PROM_PLA_RAM item in the Tools menu.

◆ Select the PLA device type and click the Next button:

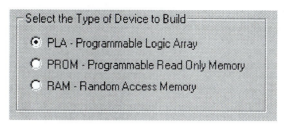

◆ Enter the desired number of inputs and outputs and select the "Enter product term data manually" data entry method. Click the Next button.

◆ For each output, enter the term data in the text box provided, using the format described earlier. The Format Help button can be used to bring up a description of the format along with some examples.

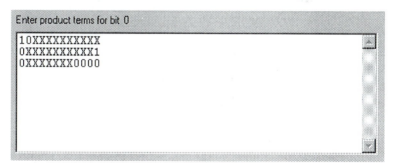

◆ Click the "Next" button.

◆ Enter the name that you wish the new part to be saved under and select the destination library. If you need to create a new library, click the New Lib button.

NOTE: We do not recommend saving your own parts to any of the standard LogicWorks release libraries. This greatly complicates upgrading to new versions of the package.

◆ Click the Finish button to save the completed part.

The part is now ready to use by double-clicking on the new item in the parts palette. You can also make any desired graphical or attribute changes to the new part using the device symbol editor.

Editing RAM, PROM, and PLA Devices

There is no way to re-enter the RAM, PROM, or PLA parameters with an existing device. Once a device definition has been created in a library, limited changes can be made using the device symbol editor.

IMPORTANT: The RAM, PROM, or PLA device definition contains structure information that *cannot* be edited after the device is created. Adding or deleting any pins using the device symbol editor will invalidate the device definition and render it useless.

The device symbol editor can be used to make the following changes to a RAM, PROM, or PLA device, if desired:

■ Any graphical changes to the symbol.

■ Pin name, visible pin number, or pin attribute changes (including pin delay and inversion).

■ Limited pin type changes (e.g., changing to open collector).

■ Part-attribute changes.

11

Device Symbol Editing

The device symbol editor (DevEditor) is a tool used to create device symbols for use on LogicWorks schematics. It provides a complete, object-oriented drawing environment with standard drawing tools, as well as specific functions tailored for symbol creation.

Starting the DevEditor

To Create a New Part

Windows—Select the New command from the File menu and choose the Device Symbol document type. This will open an empty graphics window, allowing you to create a new device type from scratch.

Macintosh—Select the DevEditor command in the Tools menu. If an existing DevEditor window is open, it will be brought to the front. If necessary, select the New Part command in the File menu.

To Edit an Existing Part

Windows—In the Parts Palette:

◆ Right-click on the part you want to edit.

◆ From the pop-up menu that is displayed, select the Edit Part command.

Macintosh—In the Parts Palette, either:

■ Click on the part you want to edit. Then pull down the File menu, open the Libraries submenu, and choose the Edit Part command. Or:

■ ⌘-click on the part you want to edit. From the pop-up menu that is displayed, select the Edit Part command.

Drawing Tools—Windows

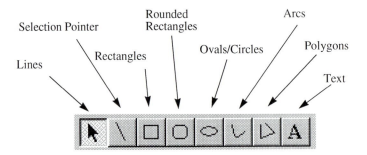

The eight items in the left half of the Tool Palette represent standard drawing tools.

Drawing Tools—Macintosh

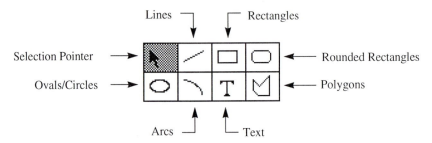

The eight items in the top half of the Tool Palette represent standard Macintosh drawing tools.

Graphic Palettes—Macintosh Only

Macintosh users can control how graphical objects are drawn by using the following pop-up palettes: fill pattern, line pattern, fill color, and line width.

If objects are selected when a palette is changed, the selected objects will be given the new characteristics. If no objects are selected when a palette is changed, the default setting for the palette will change, and objects created in the future will use the new default setting.

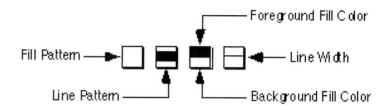

Fill Pattern Pop-Up	The Fill Pattern palette determines the pattern used to fill the inside of ovals, rectangles, and polygons. The N item refers to "No fill," which is the default.
Line Pattern Pop-Up	The Line Pattern palette sets the pattern used for all outer framing lines on objects.
Fill Color Pop-Up	The Color palette is split into upper and lower halves which respectively determine the colors used for the foreground (i.e., "black" areas of lines, text, and fill patterns) and background (i.e., "white" areas of text and fill patterns).
Line Width Pop-Up	The Line Width palette determines the width of lines and object frames.

Part Pin Tools—Windows

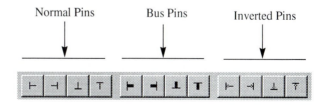

The left group of Part Pin tools are used to create normal (that is, not bus) pins on the part in any of the four orientations.

The middle group of tools are used to place bus pins on the part in any of the four orientations. Bus pins allow busses to be connected to the part. A bus pin's functionality is determined by the internal pin names it contains. See "Adding Pins" on page 252 for information on creating bus internal pins.

Tools on the right are provided as a convenience to create pins with an inversion bubble. A pin created with the inverted pin tools has no special simulation characteristics; these tools are provided only as a graphical convenience.

Part Pin Tools—Macintosh

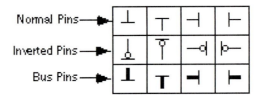

In the Macintosh version of LogicWorks, the top row of Part Pin tools are used to create pins on the part in any of the four orientations.

The second row of pins are provided as a convenience to create pins with an inversion bubble. A pin created with the inverted pin tools has no special simulation characteristics; these tools are provided only as a graphical convenience.

Pins in the third row are used to place bus pins on the part in any of the four orientations. Bus pins allow busses to be connected to the part. A bus pin's functionality is determined by the internal pin names it contains. See the section on Adding Pins, below, for information on creating bus internal pins.

Placing Pins

When placing pins, a graphical pin is associated with a name in the Pin Name List. The association is made by applying the following rules, using the first one that matches.

◆ Associate a selected pin name in the Pin Name List which is unplaced and is of the same type. For example, normal and inverted pins can't be associated with bus pin names or internal bus pin names.

◆ Associate an unplaced pin name of the correct type which follows the first selected pin name. Pin names are examined in a cyclic order, so if the bottom of the Pin Name List is reached, the search continues from the beginning.

◆ If no association is made, then a new pin name of the correct type is created and added to the bottom of the list.

Pin Name List

The Pin Name List box contains a scrollable list of the pin names associated with this device. This list is derived from the following sources:

▪ If the symbol was opened from an existing part, then the initial name list will be the pin names associated with the part.

▪ If a pin tool is clicked in the drawing area, a new name may be added with the form "PINx," where x is a sequential number. This is only done if no unplaced pin name could be used.

▪ If an open LogicWorks circuit is selected as a subcircuit, then the names of the port connectors in the circuit will be merged with the names in the Pin Name List.

▪ The Add Pins command can be used to create a list of pins to be merged with the pin names in the Pin Name List.

▪ The Autocreate Symbol command can be used to specify all the pin names and pin numbers for a device symbol.

The pin name that appears in the Pin Name List does not need to be the same as the graphical annotation that appears next to a graphical pin in the part's symbol. Some operations performed by the Autocreate Symbol command assume a correspondence, but this is done only as a convenience. The result is that changing a pin name in the Pin Name List will never cause the annotation next to a graphical pin to automatically change.

The relationship between pin names and graphical pins is not symmetrical. Every graphical pin must have a pin name, but a pin name need not have an associated graphical pin. This can lead to some surprises. For example:

■ Selecting all graphical pins in the drawing area may not select all pin names. Unplaced and internal pin names will not be selected.

■ Deselecting all graphical pins in the drawing area may not deselect all pin names. Additional unplaced or internal names may have been initially selected.

Pin List Display–Windows

Any pin that has been placed (i.e., that has a corresponding graphical pin object that appears on the symbol) will have a black " ⊣ " mark beside it. Bus pins are marked in the list with a black " ⊢ ". Pins that have not been placed on the symbol are displayed in red. Bus internal pins (pins that define the contents of a bus pin) are shown with a dotted line. Here is a typical example:

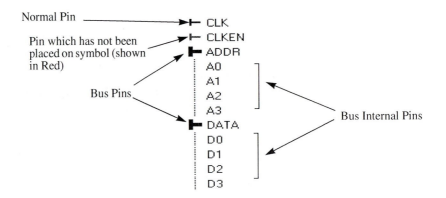

If a placed pin name (that is, a name that has a black " ⊣ " mark next to it and has an associated graphical pin) is selected, then both the pin name and the graphical name will be selected. The opposite is also true: If a graphical pin is selected, its pin name will also be selected. If a pin name is selected but has not been placed, then only the name will be selected. Bus internal pin names never have graphical associations, so selecting an internal pin name never selects a graphical pin.

Pin List Display–Macintosh

Any pin name that has been placed (i.e., that has a corresponding graphical pin object that appears as part of the symbol) will have a " ⊣ " mark beside it. Bus pins are marked in the list with "[ ]" following their names, and are displayed in bold. Internal pin names (pin names that define the contents of a bus) are indented when displayed.

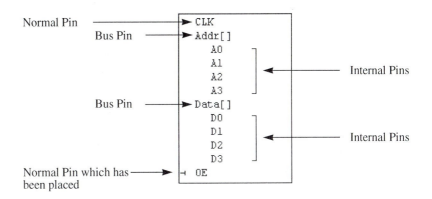

If a placed pin name (i.e., a name that has a " ⊣ " mark next to it and has an associated graphical pin) is selected, then both the pin name and the graphical name will be selected. The opposite is also true: If a graphical pin is selected, its pin name will also be selected. If a pin name is selected but has not been placed, then only the name will be selected. Internal pin names never have graphical associations, so selecting an internal pin name never selects a graphical pin.

Shift Key Usage

The (SHIFT) key on the keyboard serves two purposes in the Device Editor:

▧ Holding down the (SHIFT) key while clicking on an object adds to the current selection (that is, selects the object without deselecting the other selected objects).

▧ Holding down the (SHIFT) key while using the graphical tools may augment the operation of the selected tool.

Editing Pin Information–Windows

This section provides information on viewing and setting pin information such as names, numbers, attributes and simulation function.

Changing the Pin Name

To edit a pin name, double-click on the name in the list and edit the name in place. If there is a text annotation on the symbol labeling this pin, it will *not* automatically be updated. This must be done by editing the symbol manually.

Reordering Pins in the List

To change the order in which pins appear in the list, simply click and drag to the desired location. Multiple items can be moved in one operation by selecting items using the (SHIFT) or (CTRL) keys before dragging.

NOTE: The association between a bus pin and its internal pins is determined solely by order. A bus internal pin is always associated with the immediately preceding bus pin. Moving a bus internal pin to a position after a different bus pin will change this relationship.

Setting the Pin Number

To set a default pin number on a pin, select the pin in the pin list and enter the number in the Pin Number box at the top of the pin palette.

Setting the Pin Simulation Function

To set the simulation function of a pin (or any number of selected pins), select the desired items in the list and click on the Pin Function pop-up menu at the top of the pin palette. More information on pin functions can be found in "Appendix B— Device Pin Types" on page 405.

Deleting a Pin

To delete one or more pins from the list, simply select the desired items and press the Delete key on the keyboard.

NOTE: This cannot be undone!

Editing Pin Information–Macintosh

Showing the Pin Information Palette

To make the floating Pin Information Palette appear, double-click on any pin name in the Pin Name List. The window will display the state of the selected pin name.

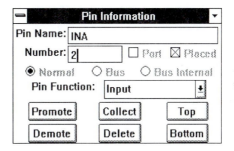

The window is divided into two parts. The top is used to examine and modify the characteristics of a pin name. If you select multiple pins in the Pin Name List and then double-click on one of them, you will narrow the selection back to one pin and display the Pin Information dialog. At this point, you can edit the pin information.

If you then select several pins in the Pin Name List by (SHIFT)-clicking, you will not be able to change any information in the dialog.

The bottom of the window is used to place, remove, or reposition pin names in the Pin Name List. All of the buttons in the lower half of the window affect all selected pin names.

Pin Information Fields and Buttons

Pin Name

The Pin name field has room for a 16-character identifier. The name must be unique within its context. The context for normal and bus pin names is all other normal and bus pin names. The context for a internal pin name is all other internal pin names in the same bus, except when the internal pin name is not yet in a bus. In this case, its context is the same as normal and bus pin names. This means that regular pin names and/or bus pin names may not be the same, but that internal pin names only need to be unique within their own bus.

If only one pin name is selected in the Pin Name List, then the name field can be changed. The change to the name will be applied when a new selection is made in the Pin Name List, or when a hard return is typed. If the name is not acceptable for some reason, a beep will occur and the name will not be changed.

If more then one pin name is selected in the Pin Name List, then the name field will show the first pin name selected, but the contents cannot be changed.

Pin Number

A four-character identifier is displayed in the pin number field. Even though this is referred to as a number, it may contain any valid character. For example, "10", "14", "E", and "In" are all valid pin numbers. The value entered for the number is displayed on the stem of a pin.

If more then one pin name is selected in the Pin Name List, then the number field will show the number of the first pin name selected, but the contents of the field cannot be changed.

Normal, Bus, and Bus Internal Buttons

The three radio buttons marked Normal, Bus, and Bus Internal show the current type of the pin name. If the pin name is not associated with a graphic pin, then the radio buttons are enabled and may be changed. Changing the radio buttons causes the pin name to be displayed differently in the Pin Name List. If the pin name is associated with a graphic pin, then the radio buttons will not be enabled but will display the current state of the pin name.

If more then one pin name is selected in the Pin Name List then the buttons will show the type of the first pin name selected, but the selected button cannot be changed.

Pin Function Drop-Down List

The Pin Function drop-down list allows the type of the pin to be specified. The list will only be displayed for pin names of type Normal or Bus Internal. Pin names of type Bus do not have their own type.

Using the drop-down list, the following functions may be assigned to the pin name: Input, Output, Tristate, Bidirectional, Open Collector, Open Emitter, Tied High, Tied Low, Latched Input, Latched Output, Clocked Input, Clocked Output, Clock Input, and No Connect.

The pin type is important to simulation—is this pin driving or driven?— and additionally, it can be important in error checking and in report and netlist generation.

◆   See Appendix B, Device Pin Types, for more information on pin types.

If more than one pin name is selected in the Pin Name List, then the Pin Function drop-down list will not be displayed.

Port Checkbox

The Port checkbox cannot be changed by the user, and is only used for information. When checked, it indicates that the pin name matches a port name in an attached subcircuit. If the box is not checked, then one of the following is true:

▩   There is no attached subcircuit.

▩   No port with the same name exists in the subcircuit.

▩   The pin is a bus internal pin and the pin name's bus name is not the same as the bus port in the subcircuit.

Placed Checkbox

The Placed checkbox cannot be changed by the user and is only used for information. When checked, it indicates that the pin name is associated with a placed graphical pin.

Pin Name List Buttons

Top and Bottom Buttons

The top and bottom buttons move the selected pin names in the Pin Name List to the top or bottom of the Pin Name List, respectively.

Promote and Demote Buttons

The promote and demote buttons move the selected pin names up or down, respectively, one position in the Pin Name List. If a selected pin name reaches either the top of the list when promoting, or the bottom of the list when demoting, then it will stop and the remaining selected pin names will begin to collect at the top or bottom of the list, respectively.

Collect Button

The collect button causes all of the selected pin names in the Pin Name List to be grouped following the first selected pin name.

Delete Button

The selected pin names in the Pin Name List and their associated graphic pins are deleted after the user is prompted to see if this is OK.

Changing the Selection

To change a displayed pin name, click on it in the Pin Name List. Once the Pin Information Palette is active (i.e., has a flashing insertion point), you can also use keyboard shortcuts. Pressing the (ENTER) key moves the selection to the next pin name and pressing (SHIFT)/(ENTER) moves the selection backward one pin name. On the Macintosh, pressing (OPTION)/(ENTER) moves

the selection to the bottom of the Pin Name List, and pressing (OPTION)/
(SHIFT)/(ENTER) moves the selection to the top of the list.

DevEditor Procedures

This section outlines step-by-step procedures for common symbol editing
tasks.

Creating a New Part

Windows—Select the New command in the File menu and choose the Device
Symbol document type.

Macintosh—Select the DevEditor command from the Tools menu on the
LogicWorks menu bar. If there were no open DevEditor windows, this will
open an empty drawing window with a default name and an empty Pin
Name List. If a DevEditor window already exists, then selecting DevEditor
in the Tools menu brings this window to the front. You can access a new
DevEditor window by selecting New Part from the File menu or by closing
all open DevEditor windows and again selecting DevEditor from the Tools
menu.

Drawing the Graphics

Draw the graphical shape that represents the part. Do this using the line,
rectangle, rounded-rectangle, oval/circle, arc, and/or polygon tools.

NOTE: Device pins *must* be added using the pin tools. Do not draw any of the
part's *pins* with the graphic tools mentioned above.

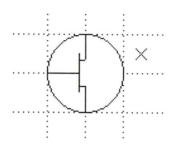

Adding Pins

Add pins to the part using the 12 pin-placement tools:

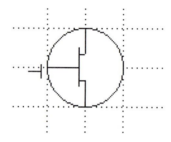

Each time you place a pin, a new default pin name is added to the Pin Name List. The pin name, number, and other information can be edited in this list. For more information on the pin list, see "Pin Name List" on page 233.

For this example, we want to name the pins "Source," "Gate," and "Drain." In addition, we want to give them the pin numbers "S," "G," and "D."

Windows—

◈ Double-click on the first pin in the list ("PIN1") and change its name to Source.

◈ While it is selected, set the pin number to the letter "S" (note that pin numbers are not restricted to decimal digits).

◈ Repeat the procedure for the seond pin, changing its name to "Gate" and pin number to "G."

◈ Repeat the procedure for the third pin, changing its name to "Drain" and pin number to "D."

Macintosh—

◆ First bring up the Pin Information Palette by double-clicking on the first pin name in the list. When placing the pins, we placed the top pin, then the bottom pin, and finally the pin on the left. Therefore, the names in the Pin Name List are in the same order.

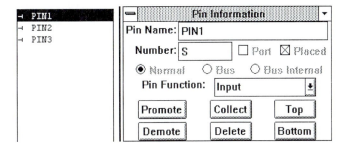

◆ For the first pin you placed, enter the name "Source" and the number "S". You can use the (TAB) key to move between the name and the number field.

◆ Press the (ENTER) key. Notice that the name was updated in the Pin Name List and that the pin number "S" has been displayed with the graphic pin.

◆ Change the next pin by clicking in its name ("PIN2") and entering the new name Gate and pin number "G" in the info palette.

◆ Press the (ENTER) key to move to the next pin.

NOTE: On the Macintosh, pressing the (RETURN) key also causes the pin information to be copied to the pin, but does not advance to the next pin name as the (ENTER) key does.

◆ Select the remaining pin and and the name "Drain" and pin number "D".

◆ When you are finished with the Pin Information Palette, you may simply close it or move it to the side.

Uses of Pin Names

The actual names given to the pins in this simple form of part are not criti-cal. However, it is a good idea to use meaningful names for the following reasons:

▨ These names will be seen in the Schematic tool's Pin Info dialog.

▨ They can be extracted in netlist output.

▨ They are used when binding pins to ports in subcircuits.

▨ In the case of bus internal pins, they are used when connecting bus pins on the part to busses.

Creating a Part with Subcircuit

To create a symbol with a subcircuit, you must perform the following steps:

◆ Create the subcircuit using the Schematic window's drawing tools. The subcircuit must have Port Connectors corresponding to the pin connections on the symbol, as described below. Leave this circuit open (i.e., displayed in a circuit window).

◆ Start the DevEditor tool, select the Subcircuit / Part Type command, and choose the "Create a subcircuit symbol and store the subcircuit with it" option. Select the subcircuit that you have just created from the list of open windows that is presented. Close the Subcircuit / Part Type window.

◆ Create the graphics for the symbol. You can do so using either the drawing tools or the Autocreate command (Windows Options menu or Macintosh DevEdit menu).

◆ Save the symbol to the library. It will be saved with a *copy* of the selected internal circuit—i.e., you can close or discard the internal circuit now, as it is saved in the library.

The following sections provide more information on this procedure.

Creating the Subcircuit

A part's subcircuit is a schematic diagram which is associated with the part. A circuit which is to be used as a subcircuit must indicate which of its signals are to correspond with the pins on the parent symbol. This is done by attaching and naming special pseudo-device symbols called port connectors. Port connectors make this association by name, i.e.: a port connector named A0 will only associate with a pin called A0 on the parent symbol.

It is important that there be an exact one-to-one match between the pins on the parent symbol and the port connectors in the subcircuit.

◆ See Chapter 6, Advanced Schematic Editing, for more information on the port interface.

Selecting the Subcircuit

The circuit to be attached to the new part must be open in a circuit window. This command will not work with a circuit file that resides on disk but is not open.

Windows—To choose the circuit to be used as the part's subcircuit, select the Subcircuit / Part Type command from the Options menu. This will display the following options box:

Macintosh—To choose the circuit to be used as the part's subcircuit, select the Subcircuit / Part Type command from the DevEdit menu. The following dialog will be displayed:

```
┌─────────────────────────────────────────────────────┐
│ Subcircuit Configurations:                          │
│   ◉ Create a subcircuit symbol, but don't store a circuit with it. │
│   ◯ Create a subcircuit symbol and store the subcircuit with it.   │
│   ◯ Import the port list from a circuit without storing the circuit. │
│   ◯ Set to Symbol Only type, never has subcircuit.  │
│   ◯ Set to Primitive Type, use with caution!        │
│                                                     │
│ Subcircuit Options:                                 │
│   ☐ Delete old pins before merging ports.           │
│   ☐ Protect part from being opened.                 │
│                                          ┌────────┐ │
│  ┌──────────────────────────────────────┤  Done  │ │
│  │                                      │└────────┘ │
│  └──────────────────────────────────────┤ Cancel │ │
│                                          └────────┘ │
└─────────────────────────────────────────────────────┘
```

In both the Windows and Macintosh versions, this box allows you to select among the following options:

Create a subcircuit symbol, but don't store a circuit with it
This is the default. It indicates that the part being made has no associated subcircuit, but doesn't rule out attaching a subcircuit (using the Attach Subcircuit command) after the symbol has been placed in a diagram.

Create a subcircuit symbol and store the subcircuit with it
When this option is selected, a selection box prompts you to choose one of the circuits that are currently open as the symbol's subcircuit. The circuit definition is included as a piece of the part and is saved with the part in the library. The circuit's port interface is imported so that the names of the ports are available in the DevEditor.

Import a port list from a circuit without storing the circuit
This selection allows you to create a symbol that has pins that match a particular subcircuit, without saving the subcircuit in the library. When this option is selected, you will be prompted to choose one of the open circuits as a subcircuit. The list of port connectors in the chosen circuit will be added to the DevEditor's pin list, but the chosen circuit *will not* be stored in the library.

Set to Symbol Only Type
This option is like the first option, except that it doesn't allow the schematic capture part of LogicWorks to associate a subcircuit with the part in the future.

Set to Primitive Type	See the section, Assigning a Primitive Type, below.

Windows—There are two subcircuit options in the lower part of the Windows dialog:

Delete pin list before merging ports	This option is only enabled when the selection made above results in a port interface being read. If this box is checked when the "Done" button is pressed, then all of the old pin names in the Pin Name List will be deleted. This allows a new port interface to be brought in without any conflicts with the existing Pin Name List. If this option wasn't checked, then port names will be merged with the names in the list. Duplicate pin names and their related properties will remain unchanged, except they are now associated with the new port. Unmatched pin names in the Pin Name List will remain exactly the same.
Subcircuit can't be opened	This option has the effect of saying, "Yes, there is a subcircuit, but in general you don't want to go into it." This causes the schematic capture part of LogicWorks to prompt for permission before entering the subcircuit.

Macintosh—There are two subcircuit options in the lower part of the Macintosh dialog:

Delete old pins before merging ports	This option is only enabled when the selection made above results in a port interface being read. If this box is checked when the "Done" button is pressed, then all of the old pin names in the Pin Name List will be deleted. This allows a new port interface to be brought in without any conflicts with the existing Pin Name List. If this option wasn't checked, then port names will be merged with the names in the list. Duplicate pin names and their related properties will remain unchanged, except they are now associated with the new port. Unmatched pin names in the Pin Name List will remain exactly the same.
Protect part from being opened	This option has the effect of saying, "Yes, there is a subcircuit, but in general you don't want to go into it." This causes the schematic capture part of LogicWorks to prompt for permission before entering the subcircuit.

Drawing the Graphics

The graphic image of the part may be drawn in the same way as was described in the previous section on creating parts without subcircuits.

◆ You can also generate a symbol automatically using the Autocreate Symbol command. See the section, Automatically Creating Symbols, below.

Associating Graphic Pins with Pin Names

In the Pin Name List, you may place any items that are shown with a red pin icon, indicating that they are unplaced. Names that have a black pin next to them already have a graphical association. Bus internal bus pins do not have a graphical representation.

You may assign graphical positions to pin names explicitly, by using the pin placement tools, or automatically, by using the Autocreate Symbol command. See "Automatically Creating Symbols" on page 254 for more information.

Explicitly assigning a graphical pin position to a pin name requires that the pin name be selected in the list and that a graphical pin be placed using one of the pin tools.

A graphical pin assignment will only be made with an unplaced pin of the correct type. If a placed pin name is selected in the Pin Name List, and an attempt is made to associate a graphical pin with it, the association will not be made because the pin name was already placed. Instead, the new graphical pin will be associated with the next free pin name of the correct type, or a new pin name will be created if there are no unplaced pin names of the correct type.

Because of the way the pin association mechanism works—that is, searching for the next placeable pin name—it is possible to place all pin names of the same type without having to make a new selection. Do this as follows:

◆ Select the first unplaced pin name and place it. Now, since the current selection is placed, the next graphic pin placed will be matched with the next pin name of the correct type.

◈ Place another pin and note how the selection moves forward to the next unplaced pin name of the same type.

Saving the Part

The newly created part is written out to any open library using the Save Part As command in the File menu.

When the menu item is selected, the Save As dialog will appear. It requires that a library be selected from the list, and that a name for the part be entered in the lower box. In this example, the Connectors library has been selected and the part's name has been left as the default "Part1."

The name you enter will become the name of the part in the library, and the title of the DevEditor window will be updated to correspond to the new part name.

NOTE: Note: Only open libraries appear in the list.

Editing an Existing Part from a Library

The following methods can be used to edit an existing part in a library.

Windows—In the Parts Palette, right-click on the part name, then select Edit Part from the pop-up menu that appears.

Macintosh—Macintosh users have three options:

■ In the Parts Palette, ⌘-click on the part name, then select Edit Part from the pop-up menu that appears. Or:

■ Just click normally on the part name in the Palette, then pull down the File menu, select the Libraries submenu, and select Edit Part. Or:

■ Open the DevEditor (by selecting it from the Tools menu), then use the Open Part command that will be enabled in the File menu.

Once opened, the new part is a *copy* of the part in the library. However, the DevEditor keeps track of the part's library of origin. If a part is selected from the library while the same part is being edited, the *original* part will be the one selected. In addition, editing a part and re-saving it to the library does *not* change all the copies of that part in all the circuits that use it.

Warnings About Editing

Observe the following cautions when editing parts.

Changing Graphics

Editing changes can be freely made to all of the part's graphical components, other than pins. It is important to remember that graphic pins are associated with pin names, and that even though graphic pins look the same, they may not be interchangeable because of their associated names.

Consider the following example: A flip flop has pins named D, CLK, and Q. The pins D and CLK are both left-facing graphic pins and Q faces to the right. Simulation and netlisting problems will occur if the graphic pins D and CLK were to have their graphical locations swapped.

Swapping the pins' positions would confuse the user because the D label would be located next to the graphical pin associated with the pin name CLK. The same would be true for the CLK pin.

To ensure that pins are placed correctly, it is important to select the pin in the Pin Name List and verify that the expected graphic pin is selected.

Changing Pin Information

If the part has a subcircuit, then it is important to remember that the pin types are derived from the ports in the subcircuit. Trying to change a pin's type without also changing the underlying subcircuit is not allowed.

For parts based on a primitive type, the functions of the pins must agree with the selected model. There is no check to ensure that the pins are correctly set for the selected model.

With other types of parts—Symbol Only and No Subcircuit—the types of the pins can be changed freely.

Changing Part Type

Changing a part's type can have dramatic effects. If you have changed a part's type, we recommend that you *not* save it into its original library with its original name. Changing the library and/or part name ensures that mistakes are easier to recover from. Some of the things to look out for are:

- Changing a subcircuit to any other type. An internal circuit is kept with the part in the library, so it is important to remember that changing the part's type will cause the circuit to be deleted. *This cannot be undone!*

- Changing to a subcircuit. When a circuit is associated with the part, its ports are read and used to create pins with the same name. These new pins carry the type of the port and will replace like-named pins that are already defined. This will result in a pin losing the type with which it was initially defined.

Saving the Part Again

A part opened from a library retains a record of its library of origin. By default, the Save Part command in the File menu will replace the original part in the library with the edited version. This operation *cannot be undone,* so unless you are sure the modified version is correct, save it to a different library or part name as described just above. This precaution allows you to test the part completely without losing the original.

Editing Part Attributes

To edit the attributes associated with a part, select the Part Attributes command in the Options menu. This will display the following dialog.

Windows—

Macintosh—

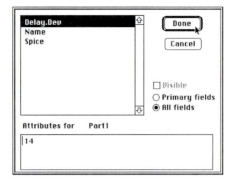

Attributes defined while in the DevEditor will be associated with the part in the library. These will become the default attributes for the part when it is placed in a circuit. Attribute values can always be changed on each copy of the device after it is placed on a schematic.

◆ For more information on how to create and modify attributes, see Chapter 6, Advanced Schematic Editing.

Adding Pins

The Add Pins command allows you to add one or more pins to the DevEditor's pin list. The pin names are created and merged with the contents of the Pin Name List when the (ENTER) key is pressed or when the Add button is clicked. The Add Pins Palette does not create graphic pins, only pin names for the Pin Name List.

The created pin names are merged with the names in the Pin Name List. If a like-named pin already exists in the list, then it may be reordered to appear in the same order as in the Add Pins Palette. If a pin number is specified, then it will replace the pin number in any existing pin with the same name.

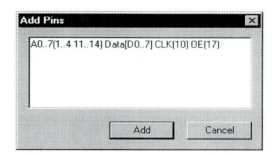

Syntax for Pin Names

Pin names may be up to 16 characters long. Three types of pin names may be described. They are:

- **Normal Pin Names** may be specified as individual names—e.g., A B C D; or as sequences—e.g., A0..3. After each pin name or pin name sequence, an optional pin number specification is allowed.

  A pin number specification defines the pin number(s) associated with the preceding pin name or pin name sequence. The specification starts with a "(" and ends with a ")." For a single pin name, the pin number specification should contain a single number. For a pin name sequence, multiple numbers and sequences may appear between the parentheses.

  In the picture above, CLK and OE both define normal pin names, which have pin numbers 10 and 17, respectively. The sequence A0..7 defines the pin names A0 through A7. These pin names were also given pin numbers. The pin numbers 1 2 3 4 11 12 13 and 14 were assigned.

  It is also possible to skip a pin name when assigning pin numbers to a pin name sequence. Consider the previous example: If we didn't want to assign the number 11 but wanted all other pin names to get the same numbers in this series, we would do the following: A0..7(1..4,,12..14) instead of: A0..7(1..4 11..14). Two commas in a row will cause a pin to be skipped when assigning the pin numbers. Three commas will cause two pins to be skipped; four commas will skip three pins; etc.

- **Bus Pin Names** are denoted by a name followed by the symbol "[...]." Bus pins need not have internal pin names defined and need not have pin numbers.

■ **Internal Pin Names** must appear between a pair of brackets: "[ ]." They have the same format as normal pin names and may have pin numbers. Internal pin names may be added, without adding a bus name pin, by not placing a bus name in front of the "[ ]." For example: [A B C] adds the internal pin names A, B, and C to the Pin Name List.

Automatically Creating Symbols

The automatic symbol creation tool will create standard rectangular symbols, given a list of the desired input and output pin names. For maximum flexibility, the generated symbol consists of separate graphic objects and is completely editable after it is generated.

Windows—Select the Auto Create Symbol command under the Options menu.

Macintosh—Select the Auto Create Symbol command under the DevEditor menu.

The current settings for line width, fill patterns, and color, and for text font, size, and style, are used in generating the symbol. The only exception to this is the part- and symbol-type name that is placed in the center of the symbol. This is rendered three points larger than the current default text setting and in bold. All text settings can be modified after the symbol is generated, if desired.

Windows—

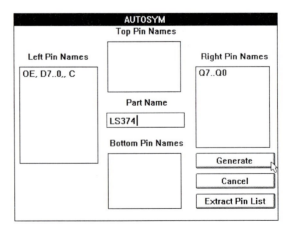

Macintosh—

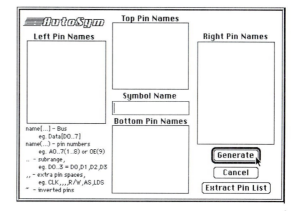

Selecting the Autocreate command displays the box shown above. The Pin Name boxes will contain the information entered the last time the DevEditor was invoked for this part. These can be modified as desired. The new settings will be merged with the Pin Name List when the Generate button is pressed.

Entering Pin Names

The four Pin Name boxes allow you to specify the names of pins to appear on the left, right, top, and bottom of the device symbol. The syntax described in the previous section, Syntax for Pin Names, is also used to define the pin names and numbers—with the following extensions:

- An inverted pin can be specified by using the ~ character in front of the name. The inverted pin will be displayed with a bubble. The ~ will not appear in the symbol.
- Items in a list can be separated by blanks or commas. Placing an extra comma between two items adds extra space between the pins on the symbol. Additional space can be added with more commas.

Entering the Part Name

The Part Name box (Windows version) or Symbol Name box (Macintosh version) allows you to specify the text that will appear centered at the top of the symbol. This also becomes the new name for the part.

Generating the Symbol

The Generate Symbol button causes the current contents of the active drawing window to be erased and replaced by the generated symbol. The Pin Name List will be merged with the new pins described in the dialog. This symbol consists of standard graphic objects, so it can be edited using any of the drawing tools provided.

An example of a device produced by the "Autocreate Symbol" command:

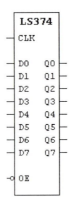

Extract Pin List Button

The Extract Pin List button updates the Pin Name boxes with the names extracted from the Pin Name List. Pins which are inputs and busses are placed on the left, outputs are placed on the right.

Assigning a Primitive Type

NOTE: The Primitive type settings should only be used with a clear understanding of their functions. Primitive types are intended primarily for use with the LogicWorks simulator. See Chapter 9, Primitive Device Types.

The primitive type of a part determines which of 38 different internal simulation models will be used to simulate the part. To describe parts which are larger or more complex, a subcircuit should be used (see the section above, Creating a Part with Subcircuit).

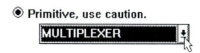

To select the primitive part type, choose the Subcircuit / Part Type command in the Options menu (Windows) or DevEdit menu (Macintosh). Then select the "Primitive" radio button. This will cause a pop-up menu to be displayed below it. The pop-up menu will show the primitive type currently selected. Clicking on the pop-up menu will display the other options available.

◈ See Chapter 9, Primitive Devices, for more information on the primitive device types available.

Creating a Gate

NOTE: This section applies only when creating a gate for use with the LogicWorks simulator. For general schematic use, use the default primitive type setting.

To make an inverter (NOT gate), place 2 pins. The order must be input, then output, … . Then set the primitive type to NOT. This will create a part with N inverters for N inputs.

To make an AND, OR, XOR, or XNOR gate, place N+1 pins. The first N pins must be inputs and the last pin an output. Then set the primitive type to the correct logical gate. This creates a single logical gate performing the logical function selected.

Creating a Breakout

NOTE: Breakouts are normally generated using the New Breakout command, described on page 289. It is possible to create custom breakouts using the DevEditor, but this is not recommended due to the possibilities for error.

A breakout is a special device that allows signals to be associated with a bus. It consists of one Bus pin, no internal pins, and N normal pins.

Because the device type is breakout, the normal pins will be connected to like-named signals in the bus.

Creating a Power (Signal) Connector

A power or signal connector is a special type of device which is generally used to represent a power or ground source—e.g.: +5V, +15V, -15V, -5V, Ground, Vss, Vdd.

These devices have special properties in the schematic. When a pin on one of these devices is connected to a signal, it attempts to assign its pin name to the signal. If the signal doesn't have a name, then it gets the name of the pin. If the signal is named and the name is different from the pin's name, then the user will be prompted to select between the signal name and the pin name.

The type of the pin must be set correctly if simulations are to make sense. For signal sources like +5V, +15V, and Vss, a normal simulation would expect a logical value of 1 (True). For signal sources like Ground and Vdd, a logical value of 0 (False) is expected. This means the pin type should be Tied High or Tied Low. (See the section, Pin Function Drop-Down List, above).

If you don't want the signal connector to supply a signal value, but only its name and the fact that it makes the signal global, then the pin type should be set to Input.

Creating a Port Connector

Port connectors have the property of associating a signal or bus with a pin on an enclosing part. The association is made by name. The port's Name is compared to the pin names on the parent part. Internal pins in busses are matched by pin name.

◆ See "Port Interface" on page 142 for more information on port connectors.

For a Signal

A port connector for a signal must have a pin that is of the correct type to interface the signal to the parent part's pin. For example, it would be correct to connect an input pin on the port connector to an output pin on the parent part. The name of the port connector pin is not important. Only the name assigned to the port connector, once it is placed, is important; it must be the same as the parent's pin name.

For a Bus

A port connector for a bus must have a bus pin that contains pins of the correct type. For example, it would be correct to connect a bus pin with three internal pins A (input), B (input), and C (output) to a parent part's bus pin that contains pins A (output), B (output), and C (input). The name of the port connector bus pin is not important, but the internal pins must have the same name. Once the port connector is placed in a circuit, its reference name is important: it must be the same as the parent's pin name.

12
Menu Reference

This chapter provides a complete guide to individual menu commands in LogicWorks. Two separate sections are provided for the Windows and Macintosh versions:

◆ For the Windows version, see "LogicWorks Menu Commands—Windows" on page 262.

◆ For the Macintosh version, see "Menu Commands—Macintosh" on page 318.

LogicWorks Menu Commands—Windows

In order to give you rapid access to commands, LogicWorks has two different sets of menus:

■ Pull-Down Menus—these are the normal File, Edit, and other menus that appear in the menu bar at the top of the application window.

■ Pop-Up Menus—these menus will appear anywhere on the schematic diagram, timing diagram, parts palette, or elsewhere, when the right mouse button is clicked. Most of the commands appearing in these menus also appear in the standard pull-down menus, so these can be considered a shortcut. The type of menu that appears depends on where the mouse is clicked. Different menus appear for Circuit, Signal, Device, Pin, and Attribute functions.

NOTE: The commands in the DevEditor menus are covered in Chapter 11, Device Symbol Editing.

LogicWorks File Menu Commands

New

This command will create a new, empty document window. Three types of documents are available:

■ Design—This choice creates a new, empty circuit window. This can be used to create an entirely new design, to create a subcircuit that will later be associated with a device, or as a temporary area to edit a circuit scrap. There is no fixed limit on the number of designs that can be open at once, although the complete contents of all open designs must fit into available memory. New Design does not create a disk file and has no effect on any files on your disk.

- Text Document—This creates a simple text document that can be used to view or edit text used by the timing, export, and PROM/PLA facilities in LogicWorks.

- Device Symbol—This creates a new device symbol document that can be used to draw a new symbol for your symbol library. Symbol editing commands are not covered in this chapter. See "Device Symbol Editing" on page 237.

Open

This command allows a design or text document to be opened from a disk file.

NOTE: Device symbols are not stored in separate files and so cannot be opened using this command. For information on opening a device symbol for editing, see "To Edit an Existing Part" on page 237.

When you open a design file, the circuit data is read into memory in its entirety and no further access to the disk file is required. LogicWorks will let you open multiple copies of the same file and will make no attempt to restrict you from writing any of them back to the same file. If you do this, it is up to you to keep track of which windows have been updated and which file you want to save each one into.

Compatibility with Older Versions

LogicWorks4 will directly read files created by older versions. However, once they have been modified and saved using this version, they can no longer be used with the older version that created them.

Close

Close closes the current document (text, design, or symbol). If the document is a design, all the Schematic windows associated with the current design are closed. If any changes have been made to your design since the last Open or Save, then you will be asked if you wish to save those changes.

NOTE: A design may be associated with multiple windows if you have been working with subcircuits. The design file is closed when the last circuit window associated with that design is closed.

Save/Save As...

Save and Save As both save the current circuit design in a circuit (.CCT) file. Save saves the circuit back into the file that it was read from. It will be disabled if no file has been opened. If you select Save As, a dialog box will be displayed requesting the name of the new file. The default name will be the current title of the circuit window (the name of the most recently opened or saved file).

Revert

This command rereads the current design from the disk file it was last saved to or read from. If any changes have been made since the last save, you will be prompted to confirm the choice before they are discarded.

Print...

Print allows you to print all or part of the current document. For design documents, if the diagram will not fit on a single page, it will be broken into as many parts as are needed, based upon the paper size specified in Print Setup. You can preview the page breaks by using the Show Printed Page Breaks option in the Design Preferences command (Schematic menu). For purposes of specifying a range to print, pages are numbered from top to bottom, then left to right. Page numbers do not appear in the printed output.

Print Setup

Presents the Print Setup dialog, which allows you to choose the size and orientation of printer paper you wish to use. Once chosen, this information will be stored with your design file and will affect the page outlines shown in the command and the Show Printed Page Breaks option in the Design Preferences command.

NOTE: LogicWorks uses the printer drivers associated with Windows. Please note that some drivers may not support features used in LogicWorks, such as rotated text.

Exit

This command exits LogicWorks.

Edit Menu Commands

Undo

This command undoes the last editing operation that was performed. The displayed name of this menu item will change based on what type of operation that was. Generally, only Schematic editing operations can be undone. Major structural changes—such as Define Attribute Fields—or any menu commands involving a dialog box are usually not undoable.

Undo never changes the contents of the Clipboard. For example, after a Cut operation, Undo will restore the Schematic, but will leave the cut object(s) on the Clipboard.

Redo

This command redoes the last Undone command. It will only be enabled immediately after an Undo operation. Any other editing operation will disable this item.

Using the Clipboard

The standard Clipboard commands Cut, Copy, and Paste can be used to move or copy circuit fragments—and graphical and text information—within a single circuit window, between multiple windows, and between different programs (e.g., word processing or drafting).

Using Clipboard Data From Other Programs

When you enter LogicWorks, the Clipboard may contain graphical or text information cut or copied from a document in another program. LogicWorks allows you to make use of this information in two ways:

■ Text information from a word processor or text editor can be pasted into a text block. See more information in "Text Objects" on page 119..

■ Graphical information copied from other applications can be pasted into LogicWorks as long as the source application supports Windows Bitmap or Windows Metafile formats. For more information on using this mechanism to create title blocks and sheet borders, see "Sheet Borders and Title Blocks" on page 121.

Using Clipboard Data From LogicWorks

When a Cut or Copy is done, three types of data are placed on the Clipboard:

■ Text data (if text labels were selected). The text can be pasted into any text editor or word processor.

■ A bitmap (.BMP format) of the selected items, which can be pasted into a graphics document in most drawing programs.

■ The LogicWorks circuit info for the selected items. This data is in a format that only LogicWorks can understand, and is discarded when you exit the program.

IMPORTANT: Circuit structural information on the Clipboard is discarded when you quit the program. Only picture and text data is retained. You cannot use the clipboard to Copy and Paste circuit data between LogicWorks sessions—you must use disk files.

Cut and Copy work on the currently selected group of circuit objects and will be disabled if no objects are selected. When items are copied onto the Clipboard, their names are copied with them; this may result in duplicate names. If duplicate signal names are pasted back into the circuit from which they were copied, then logical connections will be made between the like-named segments.

Cut

Cut removes the currently selected objects from the circuit and transfers them to the Clipboard. It is equivalent to doing a Copy and then a Delete. Cut will be disabled if no objects are selected.

Copy

Copies the currently selected objects onto the Clipboard without removing them. This can be used to duplicate a circuit group, copy it from one file to another, or to copy a picture of the circuit group to a drawing program. See the notes on Clipboard data above. Copy will be disabled if no objects are currently selected.

Paste

The Paste command, when executed in a Schematic window, replaces the mouse pointer with a flickering image of the Clipboard's contents. As noted above, this data may be a circuit group copied from within Logic-Works, or it may be text information created by another program or module. The image of the Clipboard data can be dragged around and positioned as desired. The item will be made a permanent part of your diagram when the left mouse button is pressed.

LogicWorks checks for signal connections only at "loose ends" in the signal lines being pasted—that is, at ends of line segments that do not touch devices or other line segments. For example, if the following circuit scrap was pasted, the points marked X would be checked for connection to the existing circuit.

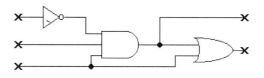

NOTE: Connection "hit testing" can be disabled by holding down the (CTRL) key while clicking the mouse button (this also applies to single device placing).

In this case, the circuit scrap is placed, but no connections will be made to adjacent items. This allows the group to be selected again (by (CTRL)-double-clicking on any device in the group) and moved without interactions with other objects in the circuit.

Paste will be disabled if the Clipboard contains no information of a recognized type.

Rotation on Paste and Duplicate

Any group of objects being Pasted or Duplicated can be rotated using the same controls as you use when placing a device:

■ The orientation tools in the toolbar.
■ The arrow keys on the keyboard.

Note that these controls are only effective while actually moving the flickering image of the object being pasted. Each Paste or Duplicate always starts in the same orientation as the source.

NOTE: The Orientation command on the Schematic menus cannot be used during Paste or Duplicate operations, because selecting this menu command will abort the paste function.

Delete

Removes the currently selected objects from the circuit. Pressing the (BACKSPACE) or (DELETE) key on the keyboard is equivalent to Delete. This command will be disabled if no devices or signals are selected.

Duplicate

Makes a copy of the selected circuit group—which can be dragged and positioned as desired. This is equivalent to selecting Copy and then Paste, except that the selected circuit scrap is not placed on the Clipboard for future use. See the notes under Paste, above, on how connections are made when a group is placed in the circuit. Note that the duplicated objects can be rotated using the orientation tools in the toolbar or the arrow keys on the keyboard.

Point

This selects the normal operating mode for LogicWorks, indicated by the arrow cursor. Selecting this command is equivalent to clicking on the ⬉ icon in the toolbar. The following functions are accessible in Point mode:

▪ By clicking on an object, you can select the object for operations using the Edit menu commands. To select an I/O device, or to select multiple objects, hold the (SHIFT) key on the keyboard while you click.

▪ By clicking and dragging near the end of a device pin or signal line, you can extend that line in any direction.

▪ By clicking and dragging a signal line anywhere except near the end, you can change its perpendicular position.

▪ By clicking and dragging any other object, you can reposition the object.

◆ All of the above functions are described in more detail in Chapter 5, Schematic Editing.

Shortcuts to Point

Since you will frequently want to return to Point mode, three shortcuts are provided for this purpose:

▪ Pressing the keyboard spacebar.

▪ Pressing the Escape key.

Text

The Text command changes the current cursor to text mode and is equivalent to clicking on the **A** icon on the toolbar. In this mode the following functions are available:

▪ A name can be associated with a device by clicking and holding on the device, then moving the cursor to the desired position for the text, then releasing the button and typing the desired name (up to 16 characters) followed by (ENTER).

■ A name can be associated with a signal by clicking and holding anywhere along a signal line, then proceeding as for devices above. Signal names differ from device names in that they can appear at multiple locations along the length of the signal line, up to a maximum of 100 positions. Additional name positions are added by simply repeating the naming procedure as many times as required. If the name at any position is altered, then all positions are updated.

■ Any existing attribute item displayed on the schematic (including a name) can be edited by clicking on it. If the text in question was being displayed with the field name or if it was rotated, then an edit box will be displayed. Otherwise it can be edited right on the diagram.

■ A pin number can be placed on a device pin by clicking on the pin within 5 pixels of the device. A blinking insertion point will appear, and you will be able to type up to 4 characters. Press the (ENTER) key to terminate the pin number.

■ Free text (i.e., text not associated with a specific device or signal) can be placed by clicking anywhere on the diagram other than on a device or signal. This text can contain hard returns or any other characters.

■ Any of the above items can be edited by clicking anywhere in the existing object. The blinking insertion point will appear in the text at the position of the click.

◆ All of the above functions are described in more detail in Chapter 5, Schematic Editing.

Zap

The Zap command changes the current cursor to Zap mode, and is equivalent to clicking on the ⌇ icon in the toolbar. When the tip of this cursor is clicked on any object in a circuit, that object is removed. Using the Zap cursor is more selective than using the Delete command on certain selected objects:

■ Signal or bus lines—The Zap tool removes only the line segment under the cursor.

■ Pin numbers—The Zap tool removes the pin number.

■ Attribute items—The Zap tool removes the visible attribute text from the diagram, but leaves the value associated with the object.

◆ See more information on this command and other editing features in Chapter 5, Schematic Editing.

Draw Signal

The Draw Signal command is equivalent to clicking on the + icon in the toolbar and places the program in signal drawing mode. In this mode you can draw or extend signal lines as follows:

■ Clicking anywhere along an existing signal line extends the signal, starting at that point.

■ A new signal can be created by clicking anywhere on the diagram.

When you click again, the lines on the screen become permanent and a new set of lines are drawn starting at that point. A number of line routing options are selected by pressing the (CTRL), (SHIFT), and/or (TAB) keys while drawing. To terminate signal drawing mode, double-click the left mouse button, then press the spacebar or click anywhere in the menu bar.

◆ See more information on signal drawing modes in Chapter 5, Schematic Editing.

Draw Bus

The Draw Bus command is equivalent to clicking on the **+** icon in the toolbar. Bus-drawing mode behaves exactly like signal-drawing mode except that a bus line is created instead of a signal line.

Select All

This command selects and highlights all elements in the current Schematic window. You can then apply Clipboard (and other) commands to the entire page.

View Menu Commands

Screen Scaling Commands

Four commands are provided which control the enlargement or reduction of the circuit diagram on the screen. These commands control screen display only, and have no effect on the stored circuit information, printed output, or graphics files. Due to the integer calculations that are done by LogicWorks and by the Windows system, device symbols and text may be displayed rather crudely at scale factors other than 100%. It is best to do most editing at normal size to ensure that everything lines up as you would expect.

Normal Size

When a circuit window is topmost, Normal Size sets the screen scale to 100%.

Reduce To Fit

Reduce to Fit sets the scale factor, and centers the display, so that the entire diagram fits in the window.

Zoom In

When a circuit window is topmost, Zoom In increases the scale factor, causing the diagram to appear larger on the screen.

Zoom Out

Zoom Out decreases the scale factor, causing the diagram to appear smaller on the screen.

Magnify

This command provides an alternative method of zooming into and out of a selected area of the diagram. When you select Magnify, the cursor changes into the Q shape.

Zooming In

Two methods of zooming in are provided:

1) Clicking and releasing the mouse button on a point on the diagram will zoom in to that point by one magnification step.

2) Clicking and dragging the mouse down and to the right zooms in on the selected area. The point at which you press the mouse button will become the top left corner of the new viewing area. The point at which you release the button will become approximately the lower right corner of the displayed area. The circuit position and scaling will be adjusted to display the indicated area.

Zooming Out

Clicking and dragging the mouse up and to the left zooms out to view more of the schematic in the window. The degree of change in the scale factor is determined by how far the mouse is moved. Moving a small distance zooms out by one step (equivalent to using the Reduce command). Moving most of the way across the window is equivalent to doing a Reduce To Fit.

Schematic Menu Commands

The Schematic menu contains commands related to drawing the schematic diagram, including viewing and setting device and signal information, positioning and scaling the drawing on the screen, and setting sheet size, display, and printing options.

Go To Selection

This command causes the circuit position and scaling to be adjusted so that the currently selected items are centered and just fit in the Schematic window. The scaling will be set to a maximum of 100%.

Orientation...

The Orientation... command sets the orientation (up, down, left, right, mirrored) that will be used next time a device is created. When this command is selected, the following dialog box is displayed:

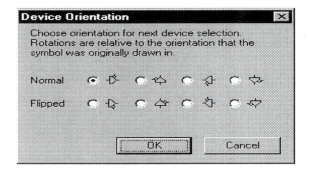

The orientation can also be changed by:

▓ Clicking directly on the orientation tools in the toolbar, or

▓ Using the arrow keys on the keyboard.

◆ See more information about symbol rotation in Chapter 5, Schematic Editing.

Get Info...

The Get Info command is a general method of viewing and setting parameters and options that are associated with the various types of objects in LogicWorks.

Showing Design and Circuit Info

If no objects are selected in the circuit (i.e., if you have clicked in an empty portion of the diagram) then Get Info will display the following general design information dialog:

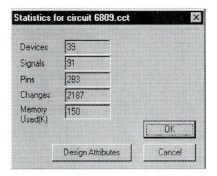

The following items of information are shown. Counts apply only to the topmost circuit level in the design, regardless of any subcircuit windows that may be open.

Devices	This is a count of devices in the selected scope. Pseudo-devices, such as ground symbols and breakouts, are not included. The count includes devices that have subcircuits.
Signals	This is a count of signal nets in the circuit, including unconnected pins.
Pins	This is a count of device pins, not including pseudo-devices.
Change	This is a count of editing changes made since the design was created. This is intended to allow comparison of different versions of the same file.
Memory used	This is a count of the main memory occupied by the selected part in the design, in Kbytes.
Design Attributes	This button displays the standard Attributes dialog for the current design.

Single Object Get Info

If a single object is selected, Get Info displays a dialog box specific to the object type. To leave any Get Info dialog, click on its OK button or press the (ENTER) or (ESC) key on the keyboard.

◆ More information on schematic objects is found in Chapter 5, Schematic Editing and Chapter 6, Advanced Schematic Editing.

General Device Info Box

When a normal device symbol is selected on the schematic (i.e., not a pseudo-device), then the following information box is displayed:

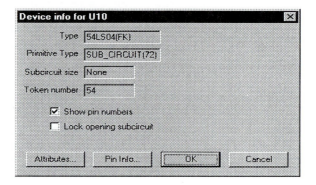

The following table lists the information and options available in this box.

Type	This is the library type name of the device symbol, i.e., the name as it appears in the Parts Palette. This is not the same as the Part attribute field, which is normally used as the part name in netlists.
Primitive Type	This is the primitive type of the symbol. For standard types, the name is shown; otherwise, the name "Reserved" is shown. The ordinal number of the primitive type value is shown in parentheses.
Subcircuit size	If the selected device has a subcircuit, its memory size is shown in Kilobytes.

Show pin numbers	This switch allows you to disable the display of pin numbers for the entire device. This is intended for discrete components or others where pin numbers are not normally shown on the diagram.
Lock opening subcircuit	This switch allows you to prevent the subcircuit (if any) of this device from being opened for editing.
Pin Info...	This button displays the Pin Information dialog (described below) for the first pin on the device. Buttons on the Pin Information dialog allow you to sequence through the other device pins.
Attributes ...	This button displays the standard Attributes dialog for the device.

NOTE: Clicking Cancel in the Device Info dialog *does not* cancel any changes that you may have made in other windows that you displayed using this dialog's option buttons.

Pseudo-Devices

If a pseudo-device is selected in the schematic, the Get Info command will be ignored.

Signal Info Box

Selecting the Get Info command with a signal selected causes the following box to be displayed:

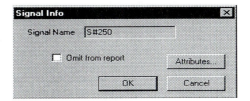

The following table describes the information and options presented in this box:

Omit from report	This checkbox controls whether the selected signal is included in any netlist output.
Attributes...	This button displays the general Attributes dialog for the selected signal.

Bus Info Box

If a bus line is selected in the schematic, the Get Info command displays the following box:

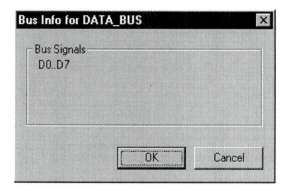

The following table describes the information and options presented in this box:

Bus Signals	This is a list of the signals contained in the bus. This list is determined by the breakouts and bus pins attached to the bus. You cannot directly change this list. See Chapter 6, Advanced Schematic Editing, for more information.
Attributes	This button displays the general Attributes dialog for the selected bus. NOTE: Bus attributes are not included in any netlist output.

General Pin Info Box

If the item selected is a device pin (non-bus and non-pseudo-device), then the following box is displayed:

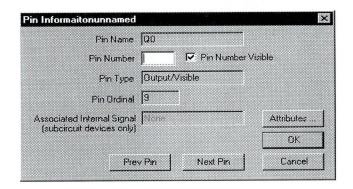

The following information and options are available:

Pin Number   This is the physical pin number corresponding to this device pin. This can be empty if desired.

Visible   This checkbox determines whether the pin number is displayed on the schematic. For some devices, such as discrete components, it may be desirable to have a pin number associated with the pin for netlist purposes without displaying it on the diagram.

Pin Type   This information item gives the function and visible status of the pin.

Pin Ordinal Number   This number is the pin's ordinal position in the device's pin list (as viewed in the DevEditor). This number can be important in some netlist formats where pin order is critical.

Associated internal signal   For subcircuit devices, this item shows the name of the signal attached to the associated port connector in the internal circuit. For other devices, this will be "None".

Attributes...   This button displays the general Attributes dialog for the selected pin.

Next Pin / Prev Pin   These buttons allow you to move to the next or previous pin (by ordinal number) on the same device, without having to return to the schematic and select the pin.

Pseudo-Device Pin Info Box

If a pseudo-device pin is selected, the Get Info command displays the signal info box for the attached signal.

Text Info Box

If a text object is selected, the Get Info command displays the following box:

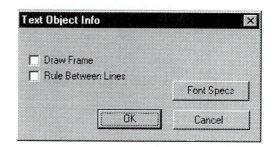

The following table summarizes the options available in this box.

Rule Between Lines   Checking this box causes a line to be drawn after each row of characters.

Draw Frame   Checking this box causes a frame to be drawn around the text item on the schematic.

Font Specs...   Clicking this button displays the standard font style dialog. Any changes made in font style affect only the selected item, but they also become the default for future text blocks.

◆   See more information on text objects in "Text Objects" on page 119.

Picture Info Box

If a pictue object is selected, the following information box is displayed:

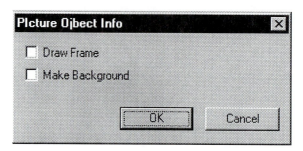

The following table summarizes the options available in this box.

Draw Frame	Checking this box causes a frame to be drawn around the picture item on the schematic.
Make Background	Selecting this option makes the picture into a background object. This means that it will not normally be selected by clicking on it. Background pictures can be selected by holding the (CTRL) and (SHIFT) keys..

◆ See more information on text objects in "Sheet Borders and Title Blocks" on page 121.

New Breakout...

The New Breakout command is used to generate a standard bus breakout symbol for a group of signals. When this command is selected, the following box will be displayed:

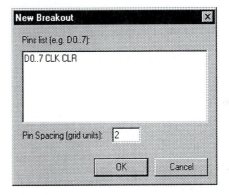

The breakout is created by entering a list of signals and the desired pin spacing, and clicking the OK button. A flickering image of the breakout will now follow your mouse movements and can be placed and connected just like any other type of device.

◆ See more information on busses and breakouts in Chapter 6, Advanced Schematic Editing.

Pin List

Type the list of desired breakout pins into this box. Rules for creating signal lists are as follows:

- Blanks or commas can be used to separate individual names in this list, therefore bussed signals cannot have names containing a blank or comma.

- A range of numbered signals can be specified using the following formats:

  D0..7 or D0..D7

  is equivalent to

  D0 D1 D2 D3 D4 D5 D6 D7

  D15..0

  is equivalent to

  D15 D14 D13 D12 D11 D10 D9 D8 E D0

  D15..D00

  is equivalent to

  D15 D14 D13 D12 D11 D10 D09 D08 D07 Es D00

  Note that the ".." format implies that bussed signal names cannot contain periods.

- The signals specified will always appear in the order given in this list from top to bottom in standard orientation. We recommend always specifying numbered signals from lowest-numbered to highest, as in the first example above, since this matches the standard library symbols.

- There is no fixed limit on the number of signals in a bus, but we recommend dividing busses up by function (i.e., address, data, control, etc.) for ease of editing.

- The same signal name can appear multiple times in the list, if desired. In this case, these pins will be connected together through the bus.

■ Any combination of arbitrarily-named signals can be included in the list, as in the following examples:

D0..15 AS* UDS* LDS*

CLK FC0..3 MEMOP BRQ0..2

Pin Spacing

The number in the Pin spacing box will be the spacing between signal pins on the breakout symbol, in grid units. The default value is 4 to match the standard LogicWorks libraries, but any number from 1 to 100 can be entered.

Push Into

This command opens the internal circuit of the given device in a separate window. This menu item will be disabled (gray) under any of the following conditions:

■ The device is not a SUBCCT (subcircuit) primitive type, or has no subcircuit.

■ The device has its "restrict open" switch set in the Device Info box.

Simply double-clicking on a device is a shortcut for the Push Into command.

NOTE: If you have used the same device type in multiple places in the design, the Push Into command creates a temporary type which is distinct from all other usages. When the subcircuit is closed, the other devices of the same type will be updated.

NOTE: See Chapter 6, Advanced Schematic Editing, for more information.

Pop Up

This command closes the current subcircuit and displays the circuit containing the parent device. If any changes have been made to the internal circuit that would affect other devices of the same type, the other devices will be updated with the new information.

◆ See more information on internal circuits and type definitions in Chapter 6, Advanced Schematic Editing.

Attach Subcircuit...

This command allows you to select an open design to attach as a subcircuit to the selected device. When this command is selected, the following dialog will appear:

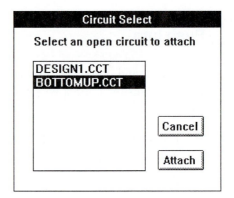

NOTE: This operation cannot be Undone!

Clicking the Attach button here will cause the following actions to be taken:

1) If the current design (i.e., the one containing the parent device) contains other devices of the same type as the selected device, then a separate, temporary type will be created for the selected device, as is done with the Push Into command, above.

2) The logical linkage between the selected device and the new internal circuit will be completed. If any mismatch is detected between the port connectors defined in the internal circuit and the pins on the parent device, you will be warned.

3) The title of the internal circuit will be updated to reflect its position in the master design.

4) The newly-attached internal circuit's window will be brought to the front. It is now considered an internal circuit that has been opened for editing and has been modified. When you close the internal circuit, you will be asked if you wish to update other devices of the same type.

Detach Subcircuit

This command turns the currently displayed subcircuit into a separate design and redefines the parent device as having no internal circuit.

IMPORTANT: 1) This operation permanently removes the subcircuit from the selected device—and from all other devices of the same type—in the selected design.

2) The Detach operation cannot be Undone!

In particular, Detach Subcircuit performs the following operations on the subcircuit displayed in the topmost window:

◆ The circuit is unlinked from its parent device, making it into a separate design.

◆ The title of the subcircuit is set to a default "Design*xxx*" name.

◆ The internal circuits of all other devices of the same type in the design are removed.

Discard Subcircuit

This command removes the subcircuit from the selected device and redefines it (and all others of the same type) as having no internal circuit.

IMPORTANT: 1) This operation permanently removes the subcircuit from the selected device—and from all other devices of the same type—in the selected design.

2) The Discard operation cannot be Undone!

Design Preferences...

The Design Preferences command is used to set a number of options which have global effect throughout a design. Selecting this command displays the following dialog:

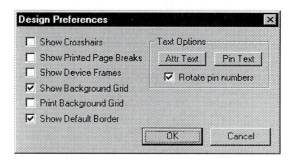

Show Crosshairs

When this option is enabled, moving crosshairs will follow all cursor movements to assist with alignment of circuit objects.

Show Printed Page Breaks

When this option is enabled, the outlines of the actual printed pages will be drawn in the circuit window. This will allow you to determine how the printer page setup will break up the circuit page for printing.

Show Device Frames

When this option is enabled, a gray outline box will be drawn around each device symbol on the schematic. This is intended to assist in locating where pins join the symbol, etc. These outlines are not shown in printed or graphics-file output.

Show Background Grid

When this option is enabled, the background grid lines will be drawn in the schematic window every 10 drawing grid units.

Print Background Grid

When this option is enabled, the background grid lines will be drawn in printed output.

Show Default Border

When this option is enabled, a border will be displayed around the boundary of the page and will adjust automatically based on printer setup.

Attribute Text Options

Clicking on the Attr Text... button will display the following font specification dialog box:

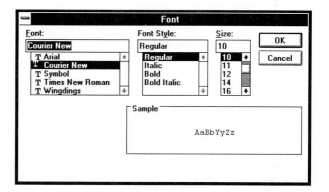

This box allows you to select the text font, size, and style used for *all* attributes displayed in the design, including:

- device and signal names
- bus breakout and bus pin labels, and
- all other attributes displayed on the diagram

…but not including:

- pin numbers (set using the Pin Text... button).

The text changes are applied when the OK button in the Design Preferences dialog is clicked. For larger designs, there may be a substantial delay while new positions of all displayed attribute items are calculated.

Pin Text Options

The Pin Text... button displays the same text specification box shown above for Attribute Text. Any changes made to this text style will be applied to all pin numbers displayed throughout the design. The text changes are applied when the OK button in the Design Preferences box is clicked.

Rotate Pin Numbers

If this item is checked, pin numbers displayed on all north- and south-facing pins will be rotated 90° to run along the length of the pin. If this item is unchecked, all pin numbers will be displayed horizontally adjacent to the pin.

Center in Page

This command on the Schematic menu moves all items on the current page so that the circuit objects (taken as a group) are centered in the page. This is intended to assist with situations where a diagram has become lopsided due to modifications.

Simulation Menu Commands

Speed

The Speed menu is used to control the simulation speed, i.e., the amount of delay inserted between simulation steps. Simulation speed can be set individually for each open circuit.

Stop

This command stops the simulation immediately. No simulation processing is done when the simulator is in this state.

Run

This command tells the simulation to proceed as fast as possible.

Other Simulation Speeds

The intermediate speed settings between Stop and Run insert various amounts of delay between executing successive simulation time steps. These can be used to slow the simulation progress for convenient observation.

Single Step

This command simulates one time step. To perform the single step, the simulator looks at the time value associated with the next signal change event in the queue, simulates the effect of that event and all following events scheduled at the same time, then returns to the stopped state. The actual time value of a single step depends on the nature of the circuit.

Simulation Params...

The Simulation Params command is a general method of setting device and pin delays and options. If no devices or pins are selected in the circuit, then this command will be disabled.

The type of dialog that is displayed will depend upon the types of devices selected, as described in the following table:

Selection	Params Box	Notes
A single Clock device	Clock Params Box	Only one clock device can be set at a time.
A single One Shot device	One Shot Params Box	Only one One Shot device can be set at a time.
Any other combination of one or more devices or pins	General Delay Box	Any selected Clock, One Shot, subcircuit, or other non-delay devices will be ignored for device delay calculations. Pin delays *can* be set on these devices.

NOTE: 1) You cannot set the device delay of a subcircuit device since its general delay characteristics are determined by its internal circuit. If any subcircuit devices are selected, they will be ignored for device delay purposes. You can set the pin delay on subcircuit devices to modify the path delay through a particular pin. Delays on the devices or pins inside a subcircuit device are not affected by any settings on the parent device using this command.

2) The Parameters command relies on numeric information in a specific format being present in the Delay.Dev or Delay.Pin attribute fields. Any invalid information in these fields will be ignored and default values used instead.

General Delay Box

For any collection of devices and pins with delay characteristics, the following box is displayed:

The controls in this box are summarized in the following table.

Devices	When this button is enabled, the other controls display and set the *device delay* characteristic of the devices currently selected in the circuit. Items such as Clock or subcircuit devices, which have no device delay characteristic, will be skipped.
Pins	When this button is enabled, the other controls display and set the *pin delay* characteristic of the pins currently selected in the circuit.
# of devices / pins	This shows a count of the devices or pins that will be affected by changes made in this box.

Shortest / longest delay   This shows the shortest and longest delays found in any of the selected devices or pins. (Note that each device or pin has only a single integer delay value associated with it.)

Delay text box   If all selected devices or pins have the same delay value, it is shown in this box. If a variety of values exist among the selected items, this box will be empty. Typing a new value (between 0 and 32,767) in this box will set all items to the given value.

+   Clicking this button will add 1 to the delays in all selected items, to a maximum value of 32,767.

–   Clicking this button will subtract 1 from the delays in all selected items, to a minimum value of zero.

1   Clicking this button will set the delay in all selected items to 1.

0   Clicking this button will set the delay in all selected items to zero.

◆ See Chapter 7, Simulation, for more information on the meaning and usage of device and pin delays.

Clock Parameters Box

When a single clock device is selected, the following parameters box is displayed:

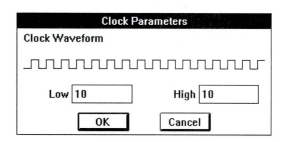

The controls in this box are summarized in the following table:

Low   This text box allows you to edit the low time setting of the selected clock device. Allowable settings are in the range 1 to 32,767.

High   This text box allows you to edit the high time setting of the
                                selected clock device. Allowable settings are in the range 1
                                to 32,767.

◆   See Chapter 9, Primitive Devices, for more information on how you
     can set the startup delay and initial value of a Clock device by setting
     the pin delay and inversion on the output pin.

One Shot Parameters Box

When a single One Shot device is selected, the following parameters box is
displayed:

```
┌──────────────────────────────────────────┐
│          One Shot Parameters             │
│ One Shot Waveform                        │
│  ─────────────────┐                      │
│                   └──────────────        │
│  ─────────────────┐ ┌──────────          │
│                   └─┘                     │
│   Delay │1        │   Width │10       │   │
│        ┌─────────┐      ┌─────────┐       │
│        │   OK    │      │ Cancel  │       │
│        └─────────┘      └─────────┘       │
└──────────────────────────────────────────┘
```

The controls in this box are summarized in the following table:

Delay   This text box allows you to edit the delay time setting of the
                                selected device. Allowable settings are in the range 1 to
                                32,767.

Width   This text box allows you to edit the width time setting of the
                                selected device. Allowable settings are in the range 1 to
                                32,767.

◆   See Chapter 9, Primitive Devices, for more information on how you
     can set the initial value of a One Shot device by setting the inversion on
     the output pin.

Add to Timing

This command adds all selected signals in the current circuit to the Timing display. If any selected items are unnamed or are already displayed, they will be ignored. New items are added at the bottom of the Timing display and will be selected after the add.

Add Automatically

When this item is checked, any signals added or edited on the schematic will automatically be added to the Timing window.

Add as Group

This command is similar to the Add to Timing command, except that the selected items are all added as a single group. Where possible, items will be sorted in alphanumeric order, with the lowest-numbered item in the least significant bit position.

Stick Signals

This command allows you to set the "stuck" status of the selected signals. It displays the following box:

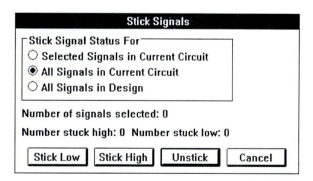

The controls in the box are summarized in the following table.

Selected Signals in Current Circuit	If this option is selected, the Stick High, Stick Low, or Unstick option will apply only to signals currently selected in the Schematic diagram.
All Signals in Current Circuit	If this option is selected, the Stick High, Stick Low, or Unstick option will apply to all signals in the circuit represented in the topmost Schematic window. Only this circuit level is affected, i.e., other open subcircuits will not be changed.
All Signals in Design	If this option is selected, the Stick High, Stick Low, or Unstick option will apply to all signals in all parts of the current design.
Number of signals selected	This displays the number of signals that will be affected by any changes made in this box.
Number stuck high	This displays the number of signals in the selected scope that are currently stuck at a high level.
Number stuck low	This displays the number of signals in the selected scope that are currently stuck low.
Stick Low	This closes the box and applies a "stuck low" value to all selected signals.
Stick High	This closes the box and applies a "stuck high" value to all selected signals.
Unstick	This unsticks all selected signals, allowing them to return to their driven value.

◆ See more information on stuck signal values in Chapter 7, Simulation.

Import Timing (Text)...

This command clears the Timing window, then opens the selected Timing text data file and pastes the data onto the diagram. This is equivalent to selecting the Clear Simulation command, then using the Paste command to place the file data at time zero. See the rules for the Paste command, below.

NOTE: This command does not display or remove any traces in the Timing window. It only reads signal event data and associates it with matching traces. If any traces are named in the file that are not currently displayed, you will be warned and that set of data will be skipped.

Export Timing (Text)...

This command saves all the displayed data in the Timing window to a text file. This file can be used for external purposes, or can be reloaded as a setup for a new simulation using the Import Timing Text command.

◆ See Appendix D, Timing Text Data Format, for a description of the file format.

Print Timing...

This command prints the contents of the Timing window using the current print setup. The current display will be divided into as many pages as required.

Print Setup...

This command determines the page setup for the Print Timing command. This can be different than the setup for the schematic diagram.

LogicWorks Help Menu

About LogicWorks...

This command displays the About LogicWorks information box.

LogicWorks Online

A number of resources are available on the World Wide Web for Logic-Works users, including technical notes, FAQs, free downloads, add-on products, and so forth. These items will direct your Web browser directly to the corresponding Web page. You must, of course, have a connection to the Internet active for these menu items to work.

Device Pop-Up Menu

A device pop-up menu is displayed by using the right mouse button to select any device in the current circuit.

Device Info...

This command displays the general Device Information box. This is equivalent to selecting the device and using the Get Info command. See more information under the Get Info command above.

Attributes...

This displays the standard Attributes dialog, allowing you to enter or edit attribute data for the selected device.

Name...

This command displays a simple edit box allowing you to enter or edit the device name. This provides a simpler method of editing only the name, as an alternative to using the Attributes command above.

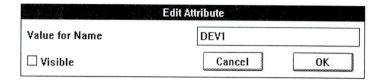

Rotate and Flip Commands

These four commands are equivalent to deleting the selected device and replacing it in the selected new orientation. If the new rotation causes any device pins to touch adjacent signal lines, connections will be made (unless the Control key is held).

◆ See more information on device rotation in "Advanced Schematic Editing" on page 125.

Cut

This is equivalent to selecting the Cut command in the Edit menu while this device is selected. The device is copied to the Clipboard and removed from the circuit.

Copy

This is equivalent to selecting the Copy command in the Edit menu while this device is selected. The device is copied to the Clipboard.

Duplicate

This is equivalent to selecting the Duplicate command in the Edit menu while this device is selected. The given device is duplicated and the program enters Paste mode immediately.

The Clipboard is not affected.

Delete

This is equivalent to selecting the Delete command in the Edit menu. The device is deleted from the circuit.

The Clipboard is not affected.

Signal Pop-Up Menu

A signal pop-up menu is displayed by using the right mouse button to select any signal line.

Signal Info...

This command displays the general signal information box. This is equivalent to selecting the signal and selecting the Get Info command in the Schematic menu. See more information on this command above.

Attributes...

This command displays the general Attributes dialog for the selected signal.

NOTE: For name changes, it is best to use the Name command described below, since it provides more options for applying the name change.

Name...

This command displays a name edit box for the selected signal. The following box is displayed, offering you two options for how to apply the changed name:

```
┌─────────────────────────────────────────────────┐
│                  Edit Attribute                  │
├─────────────────────────────────────────────────┤
│ Value for Name          │A10                    ││
│ ☐ Apply to all connected signals                 │
│ ☐ Visible                 │ Cancel │    │  OK  │  │
└─────────────────────────────────────────────────┘
```

Apply to all connected signals

This option allows you to choose whether the name change applies only to the selected signal segment (thereby breaking its connection with other like-named signals), or to all interconnected signal segments.

Visible

This option allows you to choose whether the entered name should be displayed on the schematic or not. If the name was already visible and you uncheck this box, it will be removed from the schematic. In this case, the name will still be associated with the signal as an invisible attribute. If the name was not previously visible and you check this box, it will be displayed somewhere adjacent to one of the signal line segments.

Cut

This is equivalent to selecting the Cut command in the Edit menu while this signal is selected. The signal is copied to the Clipboard and removed from the circuit.

Copy

This is equivalent to selecting the Copy command in the Edit menu while this signal is selected. The signal is copied to the Clipboard.

Duplicate

This is equivalent to selecting the Duplicate command in the Edit menu while this signal is selected. The given signal is duplicated and the program enters Paste mode immediately.

The Clipboard is not affected.

Delete

This is equivalent to selecting the Delete command in the Edit menu. The signal is deleted from the circuit.

Pin Pop-Up Menu

A pin pop-up menu is displayed by using the right mouse button to select any device pin.

NOTE: Since an unconnected device pin is both a pin and a signal, you determine whether you get the pin or signal pop-up menu as follows:

- Clicking on the pin in the last 1/4 of the pin length away from the device will display the signal menu.
- Clicking on the pin close to the device symbol will display the pin menu.

Selecting the Signal   Selecting the Pin

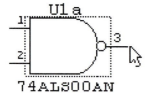
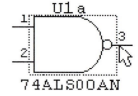

Pin Info...

This command displays a standard Pin Info box showing the name, function, and internal circuit association of the pin, as well as allowing you to edit the pin's attributes.

The Prev Pin and Next Pin buttons in this dialog can be used to view and edit other pins on the same device without having to return to the schematic and select them individually.

This command is equivalent to selecting the pin and choosing the Get Info command in the Schematic menu.

Attributes...

This command displays the standard Attributes dialog for fields associated with the selected pin.

The Prev Pin and Next Pin buttons on this box can be used to view and edit other pins on the same device without having to return to the schematic and select them individually.

Bus Pin Info...

This pop-up menu item will be enabled only when a bus pin on a device is selected. It allows the association between the bus internal pins on the device and the signals in the bus to be changed. The following box will be displayed:

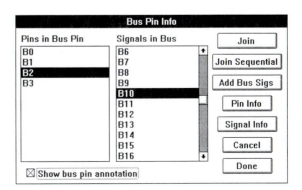

The left-hand list shows the names of the pins contained in the selected bus pin. The right-hand list shows all the signals in the attached bus. For each pin in the pin list, the signal on the same row in the signal list is the one attached to it. Signals in the signal list beyond the end of the pin list are not connected in this bus pin.

Changing Signal Connections

Two buttons are provided to change the association between pins and signals. The Join button causes the selected pin in the pin list to be joined to the selected signal in the signal list. If the selected signal is already attached to another pin in the list, then the signals will be swapped (i.e., a signal can only connect to one pin and vice versa). The signal list will be updated to show the new relationship.

The Join Sequential button provides a quick method of joining multiple numbered pins and signals. The selected pin is joined to the selected signal, as with Join, above. If the signal and pin names both have a numeric part, both numbers are incremented and the corresponding signal and pin are joined. This process is repeated until either the signal or pin name is not found in the list.

For example, given the lists appearing in the above picture, if pin B2 and signal B10 are selected, then Join Sequential would join B2–B10 and B3–B11. Since there are no more numbered pins, the process would stop. Note that although the signal and pin names are the same in this example, this is not a requirement.

Add Bus Sigs

The Add Bus Sigs button allows you to add signals to the signal list so that they can then be joined to device pins. Clicking this button displays the following box:

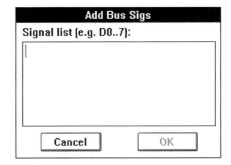

A list of signals can be typed into this box using the same format as the New Breakout command. Following are examples of allowable formats:

D0..7

D0..15 AS* UDS* LDS*

CLK FC0..3 MEMOP BRQ0..2

The order of entry will affect the order in which the signals appear in the list, but is otherwise not significant. For a complete description of the rules of this format, see the New Breakout command elsewhere in this chapter.

NOTE: These signals are only added temporarily. When you close the Bus Pin Info box, all signals that are not connected to any pin are removed from the bus.

Pin Info Button

The Pin Info button brings up the standard Pin Info box for the pin selected in the pin list. See the Get Info command for more information.

Signal Info Button

The Signal Info button brings up the standard signal info box for the signal selected in the signal list. See the Get Info command for more information.

Show bus pin annotation

If this option is enabled, a list of the connections made in the bus pin will be displayed adjacent to the pin. The format of the signal list is the format used by the Add Bus Sigs option, above.

Circuit Pop-Up Menu

Normal Size / Reduce To Fit / Zoom In/ Zoom Out

These menu items control the zoom scale factor and function exactly as the same-named items in the Schematic menu, except that they attempt to zoom in around the area of the mouse click.

Circuit Info...

This command is equivalent to the Get Info command in the Schematic menu while no items are selected in the current circuit. It displays the circuit info box.

Attribute Pop-Up Menu

When any visible attribute data item on the schematic is right-clicked, the Attribute pop-up menu will appear.

NOTE: Even though clicking on an attribute item selects and highlights the object associated with it, the commands in this menu affect only the attribute field that was clicked on.

Edit...

This command opens a text box allowing you to edit the contents of the selected field. All locations where this field is displayed on the schematic will be updated when the OK button is clicked.

Justification...

This command allows you to change the vertical and horizontal justification used in the positioning the attribute text on the diagram. When this command is selected, the following box will be displayed:

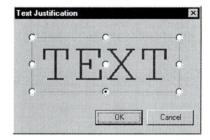

The selected point on the text is considered to be the reference point for the given attribute block. This point will be kept fixed if any field value or text style changes cause the box to be resized.

Hide

This command causes the visible attribute text that was selected to be removed from the schematic without removing the value from the field. That is, if you click on the associated object and open the Attributes dialog, the same value will still be present. If the field was displayed in more than one place, only the selected one will be removed.

Delete

This command causes the value for this field to be set to null and all visible occurrences of it on the schematic to be removed.

Duplicate

This command creates another visible occurrence of the same attribute field. This text can then be dragged or rotated to any desired position on the schematic.

Rotate Left / Rotate Right

These two commands cause the single visible attribute text item that was selected to be rotated in the given direction.

Show Field Name

This command allows you to display the field name with the value on the schematic. When this item is checked, the display will be in the form *fieldName=value*. Selecting this command again will cause the display to revert to the normal value display. This command applies only to the selected field on the selected object.

Library Manager Submenu

This menu can be displayed by clicking the right mouse button in the Parts Palette.

Edit Part

The Edit Part command opens the selected part in the device symbol editor.

NOTE: If you edit any of the parts supplied with LogicWorks, you should save the modified version to your own library. Modifying the standard libraries provided with the package is not recommended since they may be overwritten when you install an updated version, resulting in a loss of your work!

New Lib...

The New Lib... command allows you to create a new, empty symbol library file on your disk. That file will automatically be opened and will appear in the Parts Palette.

Open Lib...

The Open Lib... command allows you to select an existing symbol library file to open. The name of the library will appear in the library selection drop-down list in the Parts Palette.

NOTE: Libraries can be opened automatically when the program starts, by placing them in the default library directory or by using the LIBRARY and LIBRARYFOLDER keywords in the initialization file.

◆   See "All Libraries in a Folder" on page 414, for more information.

Close Lib...

Close Lib... allows you to close an open library and remove it from the Parts Palette. Any information required for parts used in any open designs will be automatically retained in memory. When you select the Close... command, a box will appear listing the open libraries. Pick one by clicking on it and then press the Close button, or simply double-click on the name of the library.

Lib Maintenance...

This command invokes a variety of library maintenance functions. The following box will be displayed:

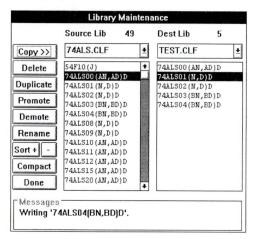

The following table summarizes the options available.

Source Lib	The Source Lib is the primary library operated on by all the command buttons. The Source Lib can be chosen from any one of the currently open libraries, by using the drop-down list at the head of the list. Any single item in the Source Lib list can be selected by clicking on it. A contiguous set of items can be selected by holding the (SHIFT) key to add to the selection.
Dest Lib	The Dest Lib is used only as the destination of the Copy command. No items can be selected in this list.
Copy	This button causes the selected parts in the Source Lib list to be copied to the Dest Lib.
Delete	This button causes the selected parts in the Source Lib to be deleted. *This cannot be undone!*
Duplicate	This causes the selected parts in the Source Lib to be duplicated—i.e., a copy of the selected items is made in the Source Lib. The Dest Lib is not affected.

Promote / Demote	The Promote and Demote buttons cause the selected items in the Source Lib to be moved up or down the list, respectively.
Rename	This button displays a box allowing a new name to be entered for the selected part.
Sort +/–	These two buttons sort the entire list in either alphabetical or inverse alphabetical order, respectively.
Compact	This button causes the Source Lib to be compacted to the destination lib (which must be empty)—i.e., any free space due to deletions is removed. See more information below.
Done	The Done button closes the Library Maintenance dialog box.

Library Compaction

When parts are deleted from a library, the free space in the file is not automatically recovered. In most cases, this is not a significant overhead. However, if a large percentage of the parts in a library have been deleted, then you may wish to compact the file. To do this:

◆ Create a new, empty library which will become the target for the Compact operation.

◆ Select the Maint command.

◆ Select the library to be compacted as the Source Lib.

◆ Select the new, empty library as the Dest Lib.

◆ Click on the Compact button.

IMPORTANT: Verify that the new destination library is correct before discarding the old copy.

Device Editor Objects Menu Commands

Bring To Front / Send To Back

These commands are used to set the front-to-back ordering of the selected objects, relative to the other graphic objects.

Group / Ungroup

The Group command causes DevEditor to treat multiple selected graphic objects—except pins—as a single object. The Ungroup command disaggregates a grouped object.

Align

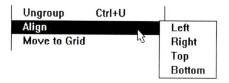

The Align submenu allows you to pick how the selected objects will be aligned. For example, Align Left causes all of the selected objects to be moved such that their left edges are aligned with the leftmost selected object's left edge.

Move to Grid

This command allows you to snap graphic objects to the grid.

Device Editor Options Menu Commands

Grids...

This command allows the user to specify the visible grid spacing, and the snap-to grids for objects drawn using the drawing tools.

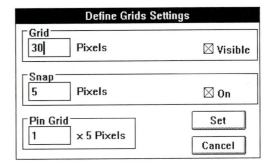

The following table summarizes the options available:

Grid Visible	This checkbox determines whether visible grid lines are shown in the drawing workspace of the DevEditor window. The spacing between these grid lines is determined by the value in the "Grid Pixels" field.
Snap On	This checkbox determines whether the corners of objects created with the drawing tools are moved to the nearest grid point.
Grid Pixels	This number determines the spacing between the visible grid lines. The measurement units are pixels at the default zoom level.
Snap Pixels	This number determines the spacing between snap-to points for the drawing tools (not including pins). This does not affect objects that have already been placed. The units are pixels at the default zoom level.
Pin Grid	This number determines the snap-to grid interval for device pins. The value will be multiplied by 5 to meet the DesignWorks pin grid requirements.

Add Pins

This command brings up the Add Pins palette allowing you to add multiple pins to the DevEditor's pin list.

◆ See Chapter 11, Device Symbol Editing, for more information.

Autocreate Symbol

This command brings up the automatic symbol creation dialog, allowing you to automatically generate a rectangular device symbol.

◆ See Chapter 11, Device Symbol Editing, for more information.

Subcircuit / Part Type

The dialog displayed when this command is selected allows you to specify the type of LogicWorks part being created. The LogicWorks types are: No Subcircuit, Subcircuit, Symbol Only, and Primitive.

◆ See Chapter 11, Device Symbol Editing, for more information.

Part Attributes

The Part Attributes... command displays the standard Attributes dialog for the part or for the selected pin, respectively. This allows you to set the default attribute values that will be used when the part is used in a schematic.

◆ See "Editing Part Attributes" on page 259 for more information.

Text Font...

This menu command displays a dialog box allowing the font, style, and size for the selected text objects to be set. If no objects are selected, then the selected text property becomes the new default.

Text Rotation

These menu items set the rotation characteristics for the selected text objects. If no objects are selected, then the text property you set becomes the new default.

Timing Trace Pop-up Menu Commands

The following commands are all associated with the Timing tool and will be available when a Timing window is topmost.

Undo

This command undoes the last editing operation in the Timing window. Unlike the Schematic tool, Timing supports only a single Undo and no Redo operation.

Copy

The Copy command copies the selected timing data to the Clipboard in picture and text format. See the notes under the Cut command, above.

Note that Copy *can* be used on a selection to the left of (older than) the current simulation time since it does not modify the selected data.

Paste

The Paste command pastes the Timing text data from the Clipboard onto the selected area of the Timing window. The following rules are used for matching the data on the Clipboard with the selected interval in the Timing window:

- Data is always pasted by name, i.e., the name of a signal in the Clipboard data will be matched to the same-named signal in the Timing window. Neither the order of the signals in the Clipboard data or the selected status of traces in the Timing window is significant. To paste data from one signal to a signal with a different name, it is necessary to paste it first into a text editor, modify the names, then paste it back.

- The Paste operation affects only signals named in the Clipboard data, regardless of the selection in the Timing window.

- The Paste operation *will not* locate signals in the schematic that are not currently displayed in the Timing window. No new traces will be added by this operation.

- If the time interval selected in the Timing window is non-zero in width, then the selected interval is deleted and all later events on pasted signals are moved forward. A time interval equal to the width of the Clipboard data is then inserted and the new data pasted into this interval.

- ◆ See more information on Timing window editing in Chapter 8, The Timing and Simulator Tools.

Select All

This command selects all traces and the entire time interval of the Timing display.

Find...

This command displays a dialog box allowing you to search for a particular signal in the Timing window.

Display On

This command enables updating of the Timing display.

Display Off

This command disables updating of the Timing display. Events are saved but are not drawn into the Timing window. This allows simulation to proceed at a substantially faster rate.

Normal Size

This command sets the horizontal display resolution to its initial defaults.

Enlarge

This command increases the horizontal display resolution in the Timing window.

Reduce

This command decreases the horizontal display resolution in the Timing window.

Timing Options...

This command displays the following dialog:

```
┌─────────────────────────────────────────┐
│            Timing Options                │
│ ┌─Timing Data Retention──────────────┐   │
│ │ ⦿ Retain displayed range only      │   │
│ │ ○ Retain for  ┌──────┐  time units  │   │
│ │               └──────┘              │   │
│ └────────────────────────────────────┘   │
│     ┌──────────┐      ┌──────────┐        │
│     │    OK    │      │  Cancel  │        │
│     └──────────┘      └──────────┘        │
└─────────────────────────────────────────┘
```

Timing Data Retention

These options allow you to determine how much signal-event data is retained in memory when a simulation is run.

Each time a signal level change occurs LogicWorks creates a record in memory containing a reference to the signal, time, new value, and source of the change. In a large simulation these records can consume enormous amounts of memory. This data can be retained for the following purposes:

▤   For use in refreshing the Timing window, should it become hidden then redisplayed.

▤   For use in timing window editing operations, such as taking the output from one circuit and using it as stimulus for another.

Data can be retained only for signals displayed in the Timing window. Signal-event data for all other signals is discarded immediately after it is no longer required for simulation.

The option "Retain displayed range only" is the normal default and results in data being discarded immediately after the corresponding point on the Timing display scrolls off the left side. This results in minimal memory usage. The setting is equivalent to entering 0 in the Retain time box.

The option "Retain for x time units" allows you to keep the signal-event data for the specified amount of time after it scrolls off the left side of the screen. If this results in a memory shortage occurring, then the simulation will stop and a message will be displayed.

Timing Label Popup Menu Commands

Get Info...

For groups, this command displays a dialog allowing reordering of the signals in the group. This affects the way the combined hexadecimal value is shown in the timing display.

For individual signals, a signal info dialog is displayed.

Go To Schematic

This command selects the signal or groups in the Schematic module corresponding to the first highlighted signal or groups in the Timing window, then brings the required Schematic window forward.

Remove

This command removes the selected signals or groups from the timing window.

Group

This command combines all the selected traces into a single display group. If any of the selected traces were already grouped, they are in effect Ungrouped first and then recombined with other selected items into a single new group.

Ungroup

This command breaks all signals in selected groups into individual traces.

Menu Commands—Macintosh

This section provides detailed information on the use of individual menu commands in LogicWorks for the Macintosh.

In order to give you rapid access to commands, the program has two different sets of menus:

▓ Pull-Down Menus—these are the normal File, Edit, and other menus that appear in the menu bar at the top of the screen.

▓ Pop-Up Menus—these menus will appear anywhere on the Schematic diagram when you hold down the ⌘ key while clicking the mouse button. Most of the commands appearing in these menus also appear in the standard menus, so they can be considered shortcuts. The type of menu that appears depends on where the mouse is clicked. Different menus appear for Circuit, Signal, Device, Pin, and Attribute functions.

◆ The commands in the DevEdit menu are covered in Chapter 11, Device Symbol Editing.

Schematic File Menu Commands

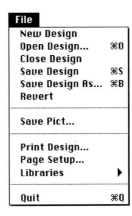

NOTE: Some of the items in the File menu may change text when other tools are in use (e.g., the DevEditor). The functions described here apply only to the Schematic tool.

New Design

This command will create a new, empty design window. This can be used to create an entirely new design, to create a subcircuit that will later be associated with a device, or as a temporary area to edit a circuit scrap.

There is no fixed limit on the number of designs that can be open at once, although the complete contents of all open designs must fit into memory. New Design does not create a disk file and has no effect on any files on your disk.

Open Design...

This command allows a design to be opened from a disk file.

When you open a file, the circuit data is read into memory in its entirety and no more access to the disk file is required. LogicWorks will let you open multiple copies of the same file and makes no attempt to restrict you from writing any of them back to the same file. If you do this, it is up to you to keep track of which windows have been updated and what file you want to save them into.

Compatibility With Older Versions

LogicWorks 3 will not directly read files created by older versions. They must first be converted using the Converter tool. For details, see the section on Installation for Macintosh in Chapter 2, Getting Started.

Close Design

Close Design closes all the circuit windows associated with the current design and removes all data from memory. If any changes have been made to your design since the last Open or Save, then you will be asked if you wish to save those changes. The same effect is achieved by clicking in the window's close box.

Save Design and Save Design As...

Save Design and Save Design As save the current circuit in a circuit file. Save Design saves the circuit back into the file that was most recently opened. It will be disabled if no file has been opened. If you select Save Design As, a dialog box will be displayed requesting the name the new file. The default name will be the current title of the circuit window, that is, the name of the most recently opened or saved file.

Revert

This command re-reads the current design from the disk file it was last saved to or read from. If any changes have been made since the last save, you will be prompted to confirm the choice before they are discarded.

Save Page Pict...

LogicWorks can save the circuit diagrams in the Macintosh standard PICT graphics format. This capability allows you to pass graphics to other programs for plotting, enhancement, or incorporation into other documentation.

PICT format is a standard format for representing graphics objects such as lines, circles, text, etc. This format can be read by most Macintosh-based graphics and drafting packages and pen-plotter drivers. PICT file format does not restrict the size of the diagram. Note that LogicWorks cannot read these files, it can only create them.

Saving the Diagram as a PICT File

Save Pict stores the circuit in the topmost circuit window to a PICT-format file. Note that variations in coordinate systems for different programs may cause the diagram to be shifted off to one side when it is read by another graphics program. This can usually be remedied by selecting everything from within the draw program and centering in the drawing space. The background grid will not appear in the file.

Plotting the Circuit Diagram

The circuit diagrams can be plotted from the PICT file using a separate plotter driver package, available from other sources.

Print Design...

Print Design allows you to print all or part of your circuit diagram. If the diagram will not fit on a single page, it will be broken into as many parts as are needed, based on the paper size specified in Page Setup. You can preview the page breaks by using the Show Page Breaks option in the Design Preferences command. For purposes of specifying a range to print, pages are numbered from top to bottom, then left to right. Page numbers do not appear in the printed output.

Laser Printer Notes

LogicWorks diagrams can be printed successfully on a LaserWriter printer (or other laser printer), but a number of precautions must be taken. In particular, problems can arise with user-created device symbols, depending on how the symbols were drawn. Bit-mapped graphics, such as those produced by "paint" type graphics programs, may not scale smoothly and should be avoided.

Page Setup

The Page Setup dialog will be presented, allowing you to choose the size and orientation of printer paper you wish to use. Once chosen, this information will be stored with your design file and affect the page outlines shown in the command and the Show Printed Page Breaks option in the Design Preferences command.

Timing File Menu Commands

The following commands are all associated with the Timing tool and will be available when a Timing window is topmost.

Open Timing Text

This command clears the Timing window, then opens the selected Timing text data file and pastes the data onto the diagram. This is equivalent to selecting the Clear Simulation command, then using the Paste command to place the file data at time zero. See the rules for the Paste command below.

NOTE: This command does not display or remove any traces in the Timing window. It only reads signal event data and associates it with matching traces. If any traces are named in the file that are not currently displayed, you will be warned and that data will be skipped.

Save Timing Text As

This command saves all the displayed data in the Timing window to a text file. This file can be used for external purposes, or can be reloaded as a setup for a new simulation using the Open Timing Text command.

◆ See Appendix D, Timing Text Data Format, for a description of the file format.

Print Timing

This command prints the contents of the Timing window using the current page setup. The current display will be divided into as many pages as required.

Page Setup

This command determines the page setup for the Print Timing command. This can be different than the setup for the Schematic diagram.

Device Editor File Menu Commands

The following commands will be available only when a Device Editor (DevEdit) window is topmost.

◆ See also Chapter 11, Device Symbol Editing, for more information on these commands.

New Part

The New Part command opens a new, empty DevEditor window.

Open Part

The Open Part command allows you to edit the part selected in the Parts Palette. If no part is selected, then a dialog will be displayed indicating that a part must be selected first. Parts may also be opened directly from the Parts Palette using the Edit Part command in the pop-up Libraries menu.

Close Part

The Close Part command closes the current DevEditor window. If any changes have been made to the open part, you will be prompted to save or discard the changes. This is the same as clicking in the DevEditor window's close box.

Save Part

The Save Part command saves the contents of the current DevEditor window back to the library it was read from. If the open part was not read from a library (i.e., it was just created), this item will be disabled.

Save Part As

The Save Part As command brings up a dialog which allows you to select an open library and the name for the part.

Libraries Submenu

```
New Lib...
Open Lib...
Close Lib...
Lib Maint...

Edit Part

Show Part Palette
```

This menu can be reached in either of the following ways:

▧ Slide down the File menu to display the Libraries submenu. (It will pop up to the right of the File menu).

▧ Hold down the ⌘ key and click anywhere in the Parts Palette.

New Lib...

The New Lib command allows you to create a new, empty symbol library file on your disk. It will be opened automatically and appear in the Parts Palette.

Open Lib...

The Open Lib command allows you to select an existing symbol library file to be opened. The name of the library will appear in the pop-up library selection menu in the Parts Palette. A small amount of memory is occupied by each open library file.

NOTE: Libraries can be opened automatically when the program starts—by placing them in the default library folder or by using the LIBRARY and LIBRARYFOLDER setup file keywords.

◆ See Chapter 2, Getting Started, for more information.

Close Lib...

Close Lib allows you to close an open library and remove it from the Parts Palette. Any information required for parts used in any open designs will be automatically retained in memory. When you select the Close Lib command, a box will appear listing the open libraries. Simply pick one by clicking on it and then press the Close Lib button, or simply double-click on the name of the library.

Edit Part

This command starts up the DevEditor tool and opens the part currently selected in the Parts Palette. This item will be disabled if nothing is selected in the Parts Palette.

Lib Maint...

This command invokes a variety of library maintenance functions. The following dialog box will be displayed:

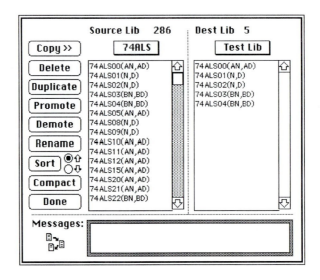

The following table summarizes the options available.

Source Lib	The Source Lib is the primary library operated on by all the command buttons. The Source Lib can be any one of the currently open libraries, selected using the pop-up menu at the head of the list. Any single item in the Source Lib can be selected by clicking on it. A contiguous set of items can be selected by holding the (SHIFT) key to add to the selection.
Dest Lib	The Dest Lib is used only as the destination of the Copy command. No items can be selected in this list.
Copy	This button causes the selected parts in the Source Lib list to be copied to the Dest Lib.
Delete	This button causes the selected parts in the Source Lib to be deleted. *This cannot be undone!*
Duplicate	This causes the selected parts in the Source Lib to be duplicated; that is, a copy of the selected items is made in the Source Lib. The Dest Lib is not affected.

Promote/Demote	The Promote and Demote buttons cause the selected items in the Source Lib to be moved up or down the list, respectively.
Rename	Rename displays a box allowing a new name to be entered for the selected part.
Sort	This button sorts the entire list alphabetically. The adjacent arrow buttons determine the direction of the sort.
Compact	This button causes the Source Lib to be compacted to the destination lib (which must be empty)—any free space due to deletions is removed. See more information below.
Done	The Done button closes the Lib Maint dialog box.

Library Compaction

When parts are deleted from a library, the free space in the file is not automatically recovered. In most cases, this is not a significant overhead. However, if a large percentage of the parts in a library have been deleted, then you may wish to compact the file. To do this:

◆ Create a new, empty library which will become the target for the Compact operation.

◆ Select the Lib Maint command.

◆ Select the library to be compacted as the Source Lib.

◆ Select the new, empty library as the Dest Lib.

◆ Click on the Compact button.

IMPORTANT: Verify that the new destination library is correct before discarding the old copy.

Schematic Edit Menu Commands

```
┌─────────────────────────┐
│ Edit                    │
│ Undo Moue          ⌘Z   │
│ Redo Moue          ⌘R   │
│ ......................  │
│ Cut                ⌘H   │
│ Copy               ⌘C   │
│ Paste              ⌘U   │
│ Clear                   │
│ Duplicate          ⌘D   │
│ ......................  │
│ Point                   │
│ Name               ⌘N   │
│ Zap                ⌘H   │
│ Draw Sig                │
│ Draw Bus                │
│ Signal Probe            │
│ ......................  │
│ Select All         ⌘A   │
│ Hide Tool Palette       │
└─────────────────────────┘
```

Undo

This command undoes the last editing operation that was performed. The text of this menu item will change based on the type of operation. Generally, only Schematic editing operations can be undone. Major structural changes (such as Define Attribute Fields), or any menu commands involving a dialog box, are usually not undoable.

Undo never changes the contents of the Clipboard. For example, after a Cut operation, Undo will restore the Schematic but will leave the Cut objects on the Clipboard.

Redo

This command redoes the last Undone command. It will only be enabled immediately after an Undo operation. Any other editing operation will disable this item.

Using the Clipboard

The standard Macintosh Clipboard commands Cut, Copy, and Paste can be used to move or copy circuit fragments—and graphical and text information—within a single circuit window, between multiple windows, and between different programs (e.g., word processing or drafting).

Using Clipboard Data From Other Programs

When you enter LogicWorks, the Clipboard may contain graphical or text information cut or copied from a document in another program. Logic-Works allows you to make use of this information in two ways:

■ Text information from a word processor or text editor can be pasted into a text block. See more information under the Text command below.

■ Picture information copied from other applications *cannot* be pasted into LogicWorks.

◆ See more information in Chapter 5, Schematic Editing.

Using Clipboard Data From LogicWorks

When a Cut or Copy is done, two types of data are placed on the Clipboard:

■ A Macintosh picture (PICT format) of the selected items, which can be pasted into a graphics document using most drawing programs.

■ The LogicWorks circuit info for the selected items. This data is in a format that only LogicWorks can understand, and is discarded when you Quit.

IMPORTANT: Circuit structural information on the Clipboard is discarded when you quit the program. Only picture and text data is retained. You cannot Copy and Paste circuit data between LogicWorks sessions—you must use disk files.

Cut and Copy work on the currently selected group of circuit objects and will be disabled if no objects are selected. When items are copied onto the Clipboard, their names are copied with them—which may result in duplicate names. If duplicate signal names are pasted back into the circuit they were copied from, then logical connections will be made between the like-named segments.

Cut

Cut removes the currently selected objects from the circuit and transfers them to the Clipboard. It is equivalent to doing a Copy and then a Clear. Cut will be disabled if no objects are selected.

Copy

Copies the currently selected objects onto the Clipboard without removing them. This can be used to duplicate a circuit group, to copy it from one file to another, or to copy a picture of the circuit group to a drawing program. See the notes on Clipboard data above. Copy will be disabled if no objects are currently selected.

Paste

The Paste command, when executed in a Schematic window, replaces the mouse pointer with a flickering image of the Clipboard's contents. As noted above, this data may be a circuit group copied from within Logic-Works, or it may be text information created by another program or module. The image of the Clipboard data can be dragged around and positioned as desired. The item will be made a permanent part of your diagram when the mouse button is pressed. Picture data from other programs *cannot* be pasted into LogicWorks.

LogicWorks checks for signal connections only at "loose ends" in the signal lines being pasted—that is, at ends of line segments that do not touch devices or other line segments. For example, if the following circuit scrap was pasted, the points marked X would be checked for connection to the existing circuit.

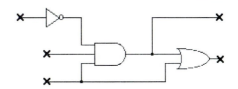

NOTE: Connection "hit testing" can be disabled by holding down the (OPTION) key while clicking the mouse button (this also applies to single device placing). In this case, the circuit scrap is placed, but no connections will be made to adjacent items. This allows the group to be selected again (by (OPTION)-double-clicking on any device in the group) and moved without interactions with other objects in the circuit.

Paste will be disabled if the Clipboard contains no information of a recognized type.

Rotation on Paste and Duplicate

Any group of objects being Pasted or Duplicated can be rotated using the same controls as you use when placing a device:

- The Orientation pop-up menu in the Tool Palette.
- The arrow keys on the keyboard.

Note that these controls are only effective while actually moving the flickering image of the object being pasted. Each Paste or Duplicate always starts in the same orientation as the source.

NOTE: The Orientation command cannot be used during Paste or Duplicate operations, since selecting this menu command will abort the paste function.

Clear

Removes the currently selected objects from the circuit. Pressing the backspace or delete key on the keyboard is equivalent to Clear. This command will be disabled if no devices or signals are selected.

Duplicate

Makes a copy of the selected circuit group—which can be dragged and positioned as desired. This is equivalent to selecting Copy and then Paste, except that the selected circuit scrap is not placed on the Clipboard for future use. See the notes under Paste, above, on how connections are made when a group is placed in the circuit. Note that the duplicated objects can be rotated using the orientation fly-out menu in the Tool Palette or the arrow keys on the keyboard.

Point

This selects the normal operating mode for LogicWorks, indicated by the arrow cursor. Selecting this command is equivalent to clicking on the ⭢ icon in the Tool Palette. The following functions are accessible in Point mode:

- By clicking on an object, you can select the object for operations using the Edit menu commands. To select an I/O device, or to select multiple objects, hold down the (SHIFT) key on the keyboard while you click.

- By clicking and dragging near the end of a device pin or signal line, you can extend that line in any direction.

- By clicking and dragging a signal line anywhere except near the end, you can change its perpendicular position.

- By clicking and dragging any other object, you can reposition the object.

◆ All of the above functions are described in more detail in Chapter 5, Schematic Editing.

Shortcuts to Point

Since you will frequently want to return to Point mode, two shortcuts are provided:

- Clicking the mouse button while pointing anywhere inside the menu bar; or

- Pressing the keyboard spacebar.

Text

The Text command changes the current cursor to Text mode and is equivalent to clicking on the A icon on the Tool Palette. In this mode the following functions are available:

■ A name can be associated with a device by clicking and holding on the device, then moving the cursor to the desired position for the text, then releasing the button, then typing the desired name (up to 16 characters) followed by (ENTER).

■ A name can be associated with a signal by clicking and holding anywhere along a signal line, then proceeding as for devices above. Signal names differ from device names in that they can appear at multiple locations along the length of the signal line, up to a maximum of 100 positions. Additional name positions are added by simply repeating the naming procedure as many times as required. If the name at any position is altered, then all positions are updated.

■ Any existing attribute item displayed on the schematic (including a name) can be edited by clicking in it. If the text in question was being displayed with the field name or if it was rotated, then an edit box will be displayed. Otherwise it can be edited right on the diagram.

■ A pin number can be placed on a device pin by clicking on the pin within five pixels of the device. A blinking insertion point will appear, and you will be able to type up to four characters. Press the (ENTER) key to terminate the pin number.

■ Free text (i.e., text not associated with a specific device or signal) can be placed by clicking anywhere on the diagram other than on a device or signal. This text can contain hard returns or any other characters.

■ Any of the above objects can be edited by clicking anywhere in the existing object. The blinking insertion point will appear in the text at the position of the click.

◆ All of the above functions are described in more detail in Chapter 5, Schematic Editing.

Zap

The Zap command changes the current cursor to Zap mode and is equivalent to clicking on the ↖ icon in the Tool Palette. When the tip of this cursor is clicked on any object in a circuit, that object is removed. Using the Zap cursor has a more selective effect on selected objects than using the Clear command:

■ **Signal or bus lines:** The Zap tool removes only the line segment under the cursor.

■ **Pin numbers:** The Zap tool removes the pin number.

■ **Attribute items:** The Zap tool removes the visible attribute text from the diagram, but leaves the value associated with the object.

◆ See more information on this command and other editing features in Chapter 5, Schematic Editing.

Draw Sig

The Draw Sig command is equivalent to clicking on the + icon in the Tool Palette and places the program in signal drawing mode. In this mode, you can draw or extend signal lines as follows:

■ Clicking anywhere along an existing signal line extends the signal, starting at that point.

■ A new signal can be created by clicking anywhere on the diagram.

When you click again, the lines on the screen become permanent, and a new set of lines are drawn starting at that point. A number of line routing options are selected by pressing the ⌘, (OPTION), and (SHIFT) keys while drawing. To terminate signal drawing mode, double-click, press the spacebar, or click anywhere in the menu bar.

◆ See more information on signal drawing modes in Chapter 5, Schematic Editing.

Draw Bus

The Draw Bus command is equivalent to clicking on the **+** icon in the Tool Palette. Bus drawing mode behaves exactly like signal drawing mode except that a bus line is created instead of a signal line.

Select All

This command selects and highlights all elements in the current circuit. You can then apply Clipboard (and other) commands to the entire page.

Show Tool Palette / Hide Tool Palette

This command hides or displays the floating Tool Palette window. You can also hide this window by clicking on its close box.

Timing Edit Menu Commands

The following commands are all associated with the Timing tool and will be available when a Timing window is topmost.

Undo

This command undoes the last editing operation in the Timing window. Unlike the Schematic tool, Timing supports only a single Undo and no Redo operation.

Cut

This command copies to the Clipboard any signal-change events on selected signals in the selected time interval, and clears the selected interval. The data is stored both in picture form and as text data. Events after the selected interval *are not* moved forward; the Delete Time command can be used to do this.

NOTE: 1) If you wish to paste a Timing picture into a word processing package, it may be necessary to first paste it into a drawing program to extract the picture data from the Clipboard. By default, a word processing package will normally take the text data from the Clipboard.

2) You cannot modify Timing data that is older than the current simulation time.

◆ See Appendix D, Timing Text Data Format, for information on the text format used to store timing data on the Clipboard.

Copy

The Copy command copies the selected Timing data to the Clipboard in picture and text format. See the notes under the Cut command, above.

Note that Copy *can* be used on a selection to the left of (older than) the current simulation time, since it does not modify the selected data.

Paste

The Paste command pastes the Timing text data from the Clipboard onto the selected area of the Timing window. The following rules are used for matching the data on the Clipboard with the selected interval in the Timing window:

▪ Data is always pasted by name, i.e., the name of a signal in the Clipboard data will be matched with the same-named signal in the Timing window. Neither the order of the signals in the Clipboard data or the selected status of traces in the Timing window is significant. To paste data from one signal to a signal with a different name, you must first paste it into a text editor, then modify the names, then paste it back.

▪ The Paste operation affects only signals named in the Clipboard data, regardless of the selection in the Timing window.

■ The Paste operation *will not* locate signals in the schematic that are not currently displayed in the Timing window. No new traces will be added by this operation.

■ If the time interval selected in the Timing window is non-zero in width, the selected interval will be deleted and all later events on pasted signals will be moved forward. A time interval equal to the width of the Clipboard data will then be inserted, and the new data will be pasted into this interval.

◆ See more information on Timing window editing in Chapter 8, The Timing and Simulator Tools.

Clear

The Clear command clears any signal-change events in the selected area in the Timing window.

NOTE: This may affect the value of a signal after the selected time interval. For example, removing a 0-to-1 transition from a signal will leave the signal in a zero state indefinitely.

Duplicate

This command inserts a duplicate of all selected signal data in the Timing window after the selected interval. In other words, the selected data is copied to a temporary location, then the selection point is moved to the end of the selected interval, and the copied data is *inserted* at this point. All signal changes after the duplicate data are moved back in time by the width of the original selection.

Point

This places the Timing window in Point mode. In this mode, the arrow cursor can be used to select data for Clipboard operations.

Draw Sig

This command places the Timing window in Draw Signal mode. In this mode, you can use the cursor to draw and edit signal-change events.

◆ Timing window editing is discussed in Chapter 8, The Timing and Simulator Tools.

Select All

This command selects all traces and the entire time interval of the timing display.

Insert Time

This command inserts a blank time interval in the selected traces. The new interval is inserted in front of the selected interval, and is of the same width as the selected interval.

Delete Time

This command deletes the selected time interval from the selected traces and moves all later data ahead by the width of the interval.

Device Editor Edit Menu Commands

The standard Edit menu commands Cut, Copy, and Paste can be used to move objects inside the DevEditor window, among LogicWorks circuit windows, and between LogicWorks and other applications. Some types of graphic objects created by other programs are not supported by the current version of DevEditor, and will not appear if pasted into the DevEditor window.

Schematic Menu Commands

The Schematic menu contains commands related to drawing the schematic diagram, including viewing and setting device and signal information, positioning and scaling the drawing on the screen, and setting sheet size, display, and printing options.

```
┌─────────────────────────┐
│ Schematic               │
├─────────────────────────┤
│ Go To Selection     ⌘G  │
│ Orientation...      ⌘J  │
│ Get Info...         ⌘I  │
│ New Breakout...     ⌘`  │
├─────────────────────────┤
│ Normal Size         ⌘=  │
│ Reduce to Fit       ⌘/  │
│ Enlarge             ⌘]  │
│ Reduce              ⌘[  │
│ Magnify             ⌘K  │
├─────────────────────────┤
│ Push Into...        ⌘,  │
│ Pop Up              ⌘Y  │
│ Attach Subcircuit...    │
│ Detach Subcircuit...    │
│ Discard Subcircuit...   │
├─────────────────────────┤
│ Design Preferences...   │
│ Center in Page          │
└─────────────────────────┘
```

Go To Selection

This command causes the circuit position and scaling to be adjusted so that the currently selected items will be centered and will just fit in the circuit window. The scaling will be set to 100% maximum.

Orientation...

The Orientation command sets the orientation (up, down, left, right, mirrored) that will be used next time a device is created. When this command is selected, the following box is displayed:

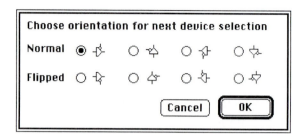

The orientation can also be changed by:

▨ Clicking directly on the orientation icon in the Tool Palette; or

▨ Using the arrow keys on the keyboard.

Since selecting any menu item cancels the device placement mode, either of the above methods is preferred since each allows you to change orientation while placing a device or circuit scrap.

◆ See more information about symbol rotation in Chapter 5, Schematic Editing.

Get Info...

The Get Info command gives you a general way to view and set parameters and options associated with the various types of objects in LogicWorks.

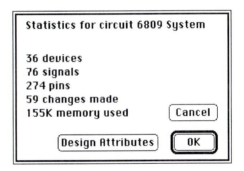 The Get Info command is also active when a Timing window is topmost and a Timing trace is selected. See more information in the section on the Timing pop-up menu, later in this chapter.

Showing Design and Circuit Info

If no objects are selected in the circuit (that is, if you have clicked in an empty portion of the diagram) then Get Info will display the following general design information box:

```
Statistics for circuit 6809 System

36 devices
76 signals
274 pins
59 changes made
155K memory used           [ Cancel ]

        [ Design Attributes ]  [[ OK ]]
```

The following items of information are shown. Counts apply only to the topmost circuit level in the design, regardless of any subcircuit windows that may be open.

N devices	This is a count of devices in the selected scope. Pseudo-devices, such as Ground symbols and breakouts, are not included. The count includes devices that have subcircuits.
N signals	This is a count of signal nets in the circuit, including unconnected pins.
N pins	This is a count of device pins, not including pseudo-devices.
N changes made	This is a count of editing changes made since the design was created. This is intended to allow comparisons of different versions of the same file.
NK memory used	This is a count of the amount of main memory occupied by the selected part of the design, in Kbytes.
Design Attributes	This button displays the standard Attributes dialog for the current design.

Single Object Get Info

If a single object is selected, Get Info displays a box specific to the object type. To leave any Get Info dialog box, click on the OK button or press the (ENTER) key on the keyboard.

◆ More information on schematic objects is found in Chapter 5, Schematic Editing, and Chapter 6, Advanced Schematic Editing.

General Device Info Box

When a normal device symbol (i.e., not a pseudo-device) is selected on the Schematic, the following information box is displayed:

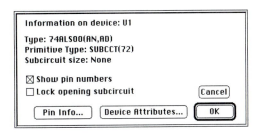

The following table lists the information and options available in this box.

Type	This is the library type name of the device symbol—that is, the name as it appears in the Parts Palette. This is *not* the same as the Part attribute field, which is normally used as the part name in netlists.
Primitive Type	This is the primitive type of the symbol. For standard types, the name is shown; otherwise, the name "Reserved" is shown. The ordinal number of the primitive type value is shown in parentheses.
Subcircuit size	If the selected device has a subcircuit, its memory size is shown in Kilobytes.
Show pin numbers	This switch allows you to disable the display of pin numbers for the entire device. This is intended for discrete components or others where pin numbers are not normally shown on the diagram.
Lock opening subcircuit	This switch allows you to prevent the subcircuit (if any) of this device from being opened for editing.
Pin Info	This button displays the Pin Info box (described below) for the first pin on the device. Buttons on the Pin Info box allow you to sequence through the other device pins.
Device Attributes	This button displays the standard Attributes dialog for the device.

NOTE: Clicking Cancel on the Device Info box *does not* cancel changes that were made in other boxes displayed using Device Info option buttons.

Pseudo-Devices

If a pseudo-device is selected in the schematic, the Get Info command will be ignored.

Signal Info Box

Selecting the Get Info command with a signal selected causes the following box to be displayed:

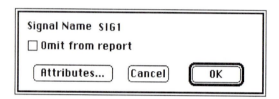

The following table describes the information and options presented in this box:

Omit from report	This checkbox controls whether the selected signal is included in any netlist output.
Attributes	This button displays the general Attributes dialog for the selected signal.

Bus Info Box

If a bus line is selected in the schematic, the Get Info command displays the following box:

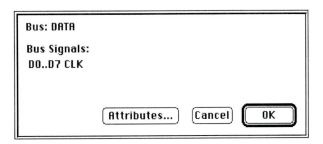

The following table describes the information and options presented in this box:

Bus Signals	This is a list of the signals contained in the bus. This list is determined by the breakouts and bus pins attached to the bus. You cannot directly change this list. See Chapter 6, Advanced Schematic Editing, for more information.
Attributes	This button displays the general Attributes dialog for the selected bus.

NOTE: Bus attributes are not included in any netlist output.

General Pin Info Box

If the item selected is device pin (non-bus and non-pseudo-device), then the following box is displayed:

```
┌─────────────────────────────────────────────────┐
│ ┌─────────────────────────────────────────────┐ │
│ │  Information on device U1 pin Y               │ │
│ │                                               │ │
│ │  Pin Number  [6        ]   ⊠ Visible          │ │
│ │                                               │ │
│ │  Pin Type: Output/Visible                     │ │
│ │  Pin Ordinal Number: 3           (Attributes) │ │
│ │  Associated internal signal: Unnamed          │ │
│ │                                   ( Cancel )   │ │
│ │       ( Next Pin )   ( Prev Pin )  (  OK  )    │ │
│ └─────────────────────────────────────────────┘ │
└─────────────────────────────────────────────────┘
```

The following information and options are available:

Pin Number	This is the physical pin number corresponding to this device pin. This can be empty if desired.
Visible	This checkbox determines whether the pin number is displayed on the schematic. For some devices, such as discrete components, it may be desirable to have a pin number associated with the pin for netlist purposes without displaying it on the diagram.
Pin Type	This information item gives the function and visible status of the pin.
Pin Ordinal Number	This number is the pin's ordinal position in the device's pin list (as viewed in the DevEditor). This can be important in some netlist formats where pin order is critical.
Associated internal signal	For subcircuit devices, this item shows the name of the signal attached to the associated port connector in the internal circuit. For other devices, this will be None.
Attributes	This button displays the general Attributes dialog for the selected pin.
Next Pin/Prev Pin	These buttons allow you to move to the next or previous pin (by ordinal number) on the same device, without having to return to the schematic and select the pin.

Pseudo-Device Pin Info Box

If a pseudo-device pin is selected, the Get Info command displays the signal info box for the attached signal.

Text Info Box

If a text object is selected, the Get Info command displays the following box:

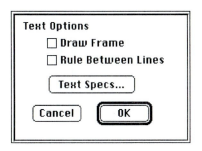

The following table summarizes the options available in this box.

◆ See more information on text objects in Chapter 5, Schematic Editing, and Chapter 6, Advanced Schematic Editing.

Draw Frame	Checking this box causes a frame to be drawn around the text item on the schematic.
Rule Between Lines	Checking this box causes a line to be drawn after each row of characters.
Text Specs	Clicking this button displays the standard text style dialog. Any changes made in text style affect only the selected item, but they also become the default for future text blocks.

New Breakout...

The New Breakout command is used to generate a standard bus breakout symbol for a group of signals. When this command is selected, the following box will be displayed:

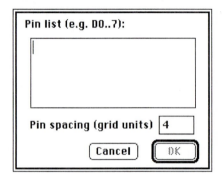

The breakout is created by entering a list of signals and the desired pin spacing, and clicking the OK button. A flickering image of the breakout will now follow your mouse movements and can be placed and connected just like any other type of device.

◆ See more information on busses and breakouts in Chapter 6, Advanced Schematic Editing.

Pin List

Type the list of desired breakout pins into this box. Rules for creating signal lists are as follows:

▨ Blanks or commas can be used to separate individual names in this list; therefore bussed signals cannot have names containing a blank or comma.

▨ A range of numbered signals can be specified using the following formats:

D0..7 or D0..D7

is equivalent to

D0 D1 D2 D3 D4 D5 D6 D7

D15..0

is equivalent to

D15 D14 D13 D12 D11 D10 D9 D8 ... D0

D15..D00

is equivalent to

D15 D14 D13 D12 D11 D10 D09 D08 D07 ... D00

Note that the ".." format implies that bussed signal names cannot contain periods.

■ The specified signals will always appear in the order given in this list from top to bottom in standard orientation. We recommend always specifying numbered signals from lowest-numbered to highest, as in the first example above, since this matches the standard library symbols.

■ There is no fixed limit on the number of signals in a bus, but we recommend dividing busses up by function (i.e., address, data, control, etc.) for ease of editing.

■ The same signal name can appear multiple times in the list, if desired. In this case, these pins will be connected together through the bus.

■ Any combination of arbitrarily named signals can be included in the list, as in the following examples:

D0..15 AS* UDS* LDS*

CLK FC0..3 MEMOP BRQ0..2

Pin Spacing

The number in the Pin Spacing box will be the spacing between signal pins on the breakout symbol, in grid units. The default value is 4 to match the standard LogicWorks libraries, but any number from 1 to 100 can be entered.

Screen Scaling Commands

Four commands are provided which control the enlargement or reduction of the circuit diagram on the screen. These commands control screen display only, and have no effect on the stored circuit information, printed output, or graphics files. The maximum enlargement allowed is 200% of normal size and the minimum reduction is 20%. Due to the integer calculations that are done by LogicWorks and by the Macintosh system, device symbols and text may be displayed rather crudely at scale factors other than 100%. It is best to do most editing at normal size to ensure that everything lines up as you would expect.

NOTE: The Normal Size, Reduce, and Enlarge commands can also be used when the Timing window is topmost. In this case they apply to the horizontal scale factor of the Timing window.

Normal Size

When a circuit window is topmost, Normal Size sets the screen scale to 100%. When a Timing window is topmost, this command sets the timing scale factor to 1 time unit per pixel.

Reduce to Fit

Reduce to Fit sets the scale factor and centers the display so that the entire diagram fits on the screen. If the size of the diagram and the size of the window are such that this would require a scale factor of less than 20%, the scale is set to 20% and the diagram is centered. If the diagram fits completely in the window at 100%, the scale is set to 100% and the diagram is centered. Reduce to Fit has no effect on the Timing window.

Enlarge

When a circuit window is topmost, Enlarge increases the scale factor by about 20%, causing the diagram to appear larger on the screen. If this causes the setting to exceed the maximum of 200%, the display is set to 200% instead. When a Timing window is topmost, Enlarge decreases the horizontal time-scale factor by one step.

Reduce

Reduce decreases the scale factor by about 20%, causing the diagram to appear smaller on the screen. If this causes the setting to go below the minimum of 20%, it is set to 20%. When a Timing window is topmost, Reduce increases the horizontal time-scale factor by one step.

Magnify

◆ This command provides an alternative method of zooming into and out of a selected area of the diagram. When you select Magnify, the cursor changes into the ⌕ shape.

Zooming In

Two methods of zooming in are provided:

▓ Clicking and releasing the mouse button on a point on the diagram will zoom in to that point by one magnification step.

▓ Clicking and dragging the mouse down and to the right zooms in on the selected area. The point at which you press the mouse button will become the top left corner of the new viewing area. The point at which you release the button will become approximately the lower right corner of the displayed area. The circuit position and scaling will be adjusted to display the indicated area.

Zooming Out

Clicking and dragging the mouse up and to the left zooms out to view more of the schematic in the window. The degree of change in the scale factor is determined by how far the mouse is moved. Moving a small distance zooms out by one step (equivalent to using the Reduce command). Moving most of the way across the window is equivalent to doing a Reduce to Fit.

Push Into...

This command opens the internal circuit of the given device in a separate window. This menu item will be disabled (gray) under either of the following conditions:

■ The device is not a SUBCCT (subcircuit) primitive type or has no subcircuit.

■ The device has its "restrict open" switch set in the Device Info box.

Simply double-clicking on a device is a shortcut for the Push Into command.

NOTE: If you have used the same device type in multiple places in the design, the Push Into command creates a temporary type which is distinct from all other usages. When the subcircuit is closed the other devices of the same type will be updated.

NOTE: See Chapter 6, Advanced Schematic Editing, for more information.

Pop Up

This command closes the current subcircuit and displays the circuit containing the parent device. If any changes have been made to the internal circuit that would affect other devices of the same type, the other devices will be updated with the new information.

◆ See more information on internal circuits and type definitions in Chapter 6, Advanced Schematic Editing.

Attach Subcircuit...

This command allows you to select an open design to attach as a subcircuit to the selected device. When this command is selected, the following box will appear:

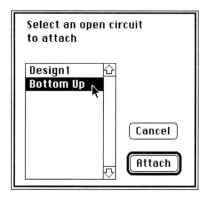

IMPORTANT: This operation cannot be undone!

Clicking the Attach button here will cause the following actions to be taken:

◆ If the current design (i.e., the one containing the parent device) contains other devices of the same type as the selected device, then a separate, temporary type will be created for the selected device, as is done with the Push Into command, above.

◆ The logical linkage between the selected device and the new internal circuit will be completed. If any mismatch is detected between the port connectors defined in the internal circuit and the pins on the parent device, you will be warned.

◆ The title of the internal circuit will be updated to reflect its position in the master design.

◆ The newly attached internal circuit's window will be brought to the front. It is now considered to be an internal circuit that has been opened for editing and modified. When you close the internal circuit, you will be asked if you wish to update other devices of the same type.

Detach Subcircuit...

This command makes the currently displayed subcircuit into a separate design and redefines the parent device as having no internal circuit.

IMPORTANT: 1) This operation *permanently* removes the subcircuit from the selected device—and from all other devices of the same type in the selected design.

2) The Detach operation cannot be Undone!

In particular, Detach Subcircuit performs the following operations on the subcircuit displayed in the topmost window:

◆ The circuit is unlinked from its parent device, making it into a separate design.

◆ The title of the subcircuit is set to a default "Design*xxx*" name.

◆ The internal circuits of all other devices of the same type in the design are removed.

Discard Subcircuit...

This command removes the subcircuit from the selected device and redefines it (and all others of the same type) as having no internal circuit.

IMPORTANT: 1) This operation *permanently* removes the subcircuit from the selected device—and from all other devices of the same type in the selected design.

2) The Discard operation cannot be Undone!

Design Preferences...

The Design Preferences command is used to set a number of options which have global effect throughout a design. Selecting this command displays the following dialog:

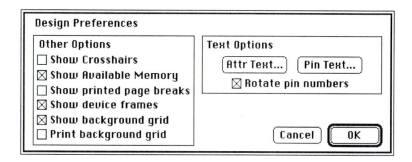

Show Crosshairs

When this option is enabled, moving crosshairs will follow all cursor movements to assist with alignment of circuit objects.

Show Available Memory

When this option is enabled, the amount of free program memory will be displayed in the schematic Tool Palette. This can be used to estimate how much room is left to expand the current design or to open additional designs.

Show Printed Page Breaks

When this option is enabled, the outlines of the actual printed pages will be drawn in the circuit window. This will allow you to determine how the printer page setup will break up the circuit page for printing.

Show Device Frames

When this option is enabled, a gray outline box will be drawn around each device symbol on the schematic. This is intended to assist in locating where pins join the symbol, etc. These outlines are not shown in printed or graphics-file output.

Show Background Grid

When this option is enabled, the background grid lines will be drawn in the schematic window every 10 drawing grid units.

Print Background Grid

When this option is enabled, the background grid lines will be drawn in printed output.

Attribute Text Options

Clicking on the Attr Text... button will display the following text specification dialog box:

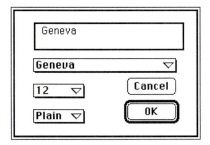

This box allows you to select the text font, size, and style used for *all* attributes displayed in the design, including:

- Device and signal names,
- Bus breakout and bus pin labels, and
- All other attributes displayed on the diagram;

    but not including:

- Pin numbers (set using the Pin Text... button).

The text changes are applied when the OK button in the Design Preferences dialog is clicked. For larger designs, there may be a substantial delay while new positions of all displayed attribute items are calculated.

Pin Text Options

The Pin Text... button displays the same text specification box shown above for Attr Text. Any changes made to this text style will be applied to all pin numbers displayed throughout the design. The text changes are applied when the OK button in the Design Preferences box is clicked.

Allow Rotated Pin Numbers

If this item is checked, pin numbers displayed on all north- and south-facing pins will be rotated 90° to run along the length of the pin. If this item is unchecked, all pin numbers will be displayed horizontally adjacent to the pin.

Center in Page

This command on the Schematic menu moves all items on the current page so that the circuit objects (taken as a group) are centered in the page. This is intended to assist with situations where a diagram has become lopsided due to modifications.

Device Pop-Up Menu

```
Device Info...
Attributes...
Name...
Push Into...
Rotate Left
Rotate Right
Flip Vertical
Flip Horizontal
..............................
Cut
Copy
Duplicate
Clear
```

A device pop-up menu is displayed by (COMMAND)-clicking on any device in the current circuit.

Device Info...

This command displays the general Device Information box. This is equivalent to selecting the device and using the Get Info command. See more information under the Get Info command above.

Attributes...

This displays the standard Attributes dialog, allowing you to enter or edit attribute data for the selected device.

Name...

This command displays a simple edit box allowing you to enter or edit the device name. This provides a simpler method of editing only the name, as an alternative to using the Attributes command above.

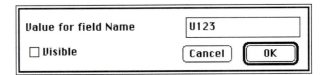

Push Into...

This command is equivalent to the Push Into command in the Schematic menu. See that entry for more details.

Flip and Rotate Commands

These four commands are equivalent to deleting the selected device and replacing it in the selected new orientation. If the new rotation causes any device pins to touch adjacent signal lines, connections will be made— unless the (OPTION) key is held down.

The Clipboard is not affected by these commands.

◆ See more information on device rotation in Chapter 6, Advanced Schematic Editing.

Cut

This is equivalent to selecting the Cut command in the Edit menu while this device is selected. The device is copied to the Clipboard and removed from the circuit.

Copy

This is equivalent to selecting the Copy command in the Edit menu while this device is selected. The device is copied to the Clipboard.

Duplicate

This is equivalent to selecting the Duplicate command in the Edit menu while this device is selected. The given device is duplicated and the program enters Paste mode immediately.

The Clipboard is not affected.

Clear

This is equivalent to selecting the Clear command in the Edit menu. The device is deleted from the circuit.

The Clipboard is not affected.

Signal Pop-Up Menu

```
┌─────────────────────────┐
│ Signal Info...          │
│ Name...                 │
│·························  │
│ Cut                     │
│ Copy                    │
│ Duplicate               │
│ Clear                   │
└─────────────────────────┘
```

A signal pop-up menu is displayed by ⌘-clicking on any signal line.

Signal Info...

This command displays the general signal information box. This is equivalent to selecting the signal and selecting the Get Info command in the Schematic menu. See more information on this command above.

Attributes...

This command displays the general Attributes dialog for the selected signal.

NOTE: For name changes, it is best to use the Name command described below, since it provides more options for applying the name change.

Name...

This command displays a name edit box for the selected signal. The following box is displayed, offering you several options for how to apply the changed name:

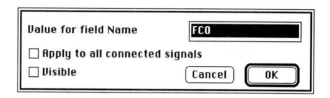

Apply to All Connected Signals

This option allows you to choose whether the name change applies only to the selected signal segment (thereby breaking its connection with other like-named signals), or to all interconnected signal segments.

Visible

This option allows you to choose whether the entered name should be displayed on the schematic. If the name was already visible and you uncheck this box, it will be removed from the schematic. In this case, the name will still be associated with the signal as an invisible attribute. If the name was not previously visible and you check this box, it will be displayed somewhere adjacent to one of the signal line segments.

Cut

This is equivalent to selecting the Cut command in the Edit menu while this signal is selected. The signal is copied to the Clipboard and removed from the circuit.

Copy

This is equivalent to selecting the Copy command in the Edit menu while this signal is selected. The signal is copied to the Clipboard.

Duplicate

This is equivalent to selecting the Duplicate command in the Edit menu while this signal is selected. The given signal is duplicated and the program enters Paste mode immediately.

The Clipboard is not affected.

Clear

This is equivalent to selecting the Clear command in the Edit menu. The signal is deleted from the circuit.

Pin Pop-Up Menu

```
Pin Info...
Attributes...
Bus Pin Info...
Push Into...
```

A pin pop-up menu is displayed by ⌘-clicking on any device pin.

NOTE: Since an unconnected device pin is both a pin and a signal you determine whether you get the pin or signal pop-up menu as follows:

- ▪ Clicking on the pin in the last 1/4 of the pin length away from the device will display the signal menu.
- ▪ Clicking on the pin close to the device symbol will display the pin menu.

Selecting the Signal   **Selecting the Pin**

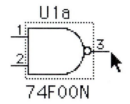   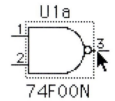

Pin Info...

This command displays a standard Pin Info box showing the name, function, and internal circuit association of the pin, as well as allowing you to edit the pin's attributes.

The Next and Previous buttons in this dialog can be used to view and edit other pins on the same device without having to return to the schematic and select them individually.

This command is equivalent to selecting the pin and choosing the Get Info command in the Schematic menu.

Attributes...

This command displays the standard Attributes dialog for fields associated with the selected pin.

The Next and Previous buttons in this dialog can be used to view and edit other pins on the same device without having to return to the schematic and select them individually.

Bus Pin Info...

This pop-up menu item will be enabled only when a bus pin on a device is selected. It allows the association between the bus internal pins on the device and the signals in the bus to be changed. The following box will be displayed:

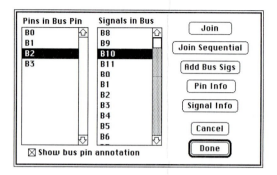

The left-hand list shows the names of the pins contained in the selected bus pin. The right-hand list shows all the signals in the attached bus. For each pin in the pin list, the signal on the same row in the signal list is the one attached to it. Signals in the signal list beyond the end of the pin list are not connected in this bus pin.

Changing Signal Connections

Two buttons are provided to change the association between pins and signals. The Join button causes the selected pin in the pin list to be joined to the selected signal in the signal list. If the selected signal is already attached to another pin in the list, the signals will be swapped—that is, a signal can only connect to one pin and vice versa. The signal list will be updated to show the new relationship.

The Join Sequential button provides a quick method of joining multiple numbered pins and signals. The selected pin is joined to the selected signal, as with Join, above. If the signal and pin names both have a numeric part, both numbers are incremented and the corresponding signal and pin are joined. This process is repeated until either the signal or pin name is not found in the list.

For example, given the lists appearing in the above picture, if pin B0 and signal B4 are selected, then Join Selected would join B0–B4, B1–B5, B2–B6, and B3–B7. Since there are no more numbered pins, the process would stop. Note that although the signal and pin names are the same in this example, this is not a requirement.

Add Bus Sigs

The Add Bus Sigs button allows you to add signals to the signal list so that they can then be joined to device pins. Clicking this button displays the following box:

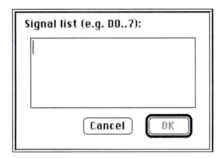

A list of signals can be typed into this box using the same format as the New Breakout command. Following are examples of allowable formats:

D0..7

D0..15 AS* UDS* LDS*

CLK FC0..3 MEMOP BRQ0..2

The order of entry will affect the order in which the signals appear in the list, but is otherwise not significant. For a complete description of the rules of this format, see the New Breakout command elsewhere in this chapter.

NOTE: These signals are only added temporarily. When you close the Bus Pin Info box, all signals that are not connected to any pin are removed from the bus.

Pin Info Button

The Pin Info button brings up the standard Pin Info box for the pin selected in the pin list. See the Get Info command for more information.

Signal Info Button

The Signal Info button brings up the standard signal info box for the signal selected in the signal list. See the Get Info command for more information.

Show Bus Pin Annotation

If this option is enabled, a list of the connections made in the bus pin will be displayed adjacent to the pin. The format of the signal list is the format used by the Add Bus Sigs option, above.

Push Into...

If the device that the pin is attached to has a subcircuit then this command will open the subcircuit and highlight the corresponding port connector.

Circuit Pop-Up Menu

```
Normal Size
Reduce to Fit
Enlarge
Reduce
..............................................
Pop Up
Circuit Info...
```

Normal Size / Reduce to Fit / Enlarge / Reduce

These menus items control the zoom scale factor and function exactly as the same-named items in the Schematic menu, except that they attempt to zoom in around the area of the mouse click.

Pop Up

This command is equivalent to the same-named item in the Schematic menu. It closes the current circuit and displays the circuit containing the parent device. It will be disabled if the current circuit is the top level in the design.

Circuit Info

Executing this command is equivalent to selecting the Get Info command in the Schematic menu while no items are selected in the current circuit. It displays the circuit info box.

Attribute Pop-Up Menu

```
Edit...
Justification...
Hide
Clear
Duplicate
Rotate Left
Rotate Right
Show Field Name
```

When you ⌘-click any visible attribute data item on the schematic, the Attribute pop-up menu will appear.

NOTE: Even though clicking on an attribute item selects and highlights the object associated with it, the commands in this menu affect only the attribute field that was clicked on.

Edit...

This command opens a text box allowing you to edit the contents of the selected field. All locations where this field is displayed on the schematic will be updated when the OK button is clicked.

Justification...

This command allows you to change the vertical and horizontal justification used in the positioning the attribute text on the diagram. When this command is selected, the following box will be displayed:

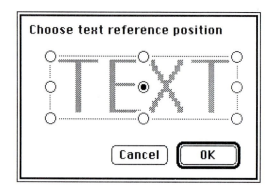

The selected point on the text is considered to be the reference point for the given attribute block. This point will be kept fixed if any field value or text style changes cause the box to be resized.

Hide

This command causes the visible attribute text that was selected to be removed from the schematic without removing the value from the field. In other words, if you click on the associated object and open the Attributes dialog, the same value will still be present. If the field was displayed in more than one place, only the one that you clicked on will be removed.

Clear

This command causes the value for this field to be set to null and all visible occurrences of it on the schematic to be removed.

Duplicate

This command creates another visible occurrence of the same attribute field. This text can then be dragged or rotated to any desired position on the schematic.

Rotate Left / Rotate Right

These two commands cause the single visible attribute text item that was selected to be rotated in the given direction.

Show Field Name

This command allows you to display the field name with the value on the schematic. When this item is checked, the display will be in the form *fieldName=value*. Selecting this command again will cause the display to revert to the normal value display. This command applies only to the selected field on the selected object.

Simulate Menu: Core Simulation Commands

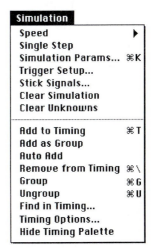

The Simulate menu contains commands for controlling simulation and Timing window functions, and for setting device simulation characteristics. If the Timing tool is not installed, the commands affecting the timing window will not appear.

Speed

The Speed submenu is used to control the simulation speed—that is, the amount of delay inserted between simulation steps. Simulation speed can be set individually for each open circuit.

Stop

This command stops the simulation immediately. No simulation processing is done when the simulator is in this state.

Run

This command tells the simulation to proceed as fast as possible.

Other Simulation Speeds

The intermediate speed settings between Stop and Run insert various amounts of delay between executing successive simulation time steps. These can be used to slow the simulation progress for convenient observation.

Single Step

This command simulates one time step. To perform the single step, the simulator looks at the time value associated with the next signal-change event in the queue, simulates the effect of that event and all following events scheduled at the same time, then returns to the stopped state. The actual time value of a single step depends on the nature of the circuit.

Simulation Params

The Simulation Params command provides a general method of setting device and pin delays and options. The type of dialog that this command displays will depend upon the types of devices selected, as described in the following:

Selection	Params Box	Notes
A single Clock device	Clock Params Box	Only one clock device can be set at a time.
A single One Shot device	One Shot Params Box	Only one One Shot device can be set at a time.
Any other combination of one or more devices or pins	General Delay Box	Any selected Clock, One Shot, subcircuit, or other non-delay device will be ignored for device delay calculations. Pin delays *can* be set on these devices.

If no devices or pins are selected in the circuit, the Simulation Params command will be disabled.

NOTE: 1) You cannot set the device delay of a subcircuit device, since its general delay characteristics are determined by its internal circuit. If any subcircuit devices are selected, they will be ignored for device delay purposes. You *can* set the pin delay on subcircuit devices to modify the path delay through a particular pin. Delays on the devices or pins inside a subcircuit device are not affected by any settings applied to the parent device using this command.

2) The Simulation Params command relies on numeric information in a specific format being present in the Delay.Dev or Delay.Pin attribute fields. Any invalid information in these fields will be ignored and default values used instead.

General Delay Box

For any collection of devices and pins with delay characteristics, the following box is displayed:

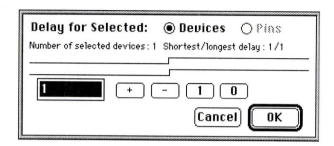

The controls in this box are summarized in the following table.

Devices	When this button is enabled, the other controls display and set the *device delay* characteristic of the devices currently selected in the circuit. Items such as Clock or subcircuit devices, which have no device delay characteristic, will be skipped.
Pins	When this button is enabled, the other controls display and set the *pin delay* characteristic of the pins currently selected in the circuit.
Number of selected devices/pins	This shows a count of the devices or pins that will be affected by changes made in this box.
Shortest/Longest delay	This shows the shortest and longest delays found in any of the selected devices or pins. (Note that each device or pin has only a single integer delay value associated with it.)
Delay Text Box	If all selected devices or pins have the same delay value, it is shown in this box. If a variety of values exist among the selected items, this box will be empty. Typing a new value (between 0 and 32,767) in this box will set all items to the given value.
+	Clicking this button will add 1 to the delays in all selected items, to a maximum value of 32,767.
–	Clicking this button will subtract 1 from the delays in all selected items, to a minimum value of zero.

| 1 | Clicking this button will set the delay in all selected items to 1. |
| 0 | Clicking this button will set the delay in all selected items to zero. |

◆ See Chapter 7, Simulation, for more information on the meaning and usage of device and pin delays.

Clock Params Box

When a single clock device is selected, the following parameters dialog is displayed:

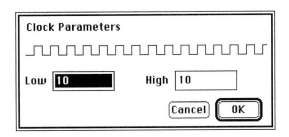

The controls in this dialog are summarized in the following table.

| **Low** | This text box allows you to edit the low time setting of the selected clock device. Allowable settings are in the range 1 to 32,767. |
| **High** | This text box allows you to edit the high time setting of the selected clock device. Allowable settings are in the range 1 to 32,767. |

◆ See Chapter 9, Primitive Devices, for more information on how you can set the startup delay and initial value of a Clock device by setting the pin delay and inversion on the output pin.

One Shot Params Box

When a single One Shot device is selected, the following parameters box is displayed:

```
┌─────────────────────────────────────────┐
│ One Shot Parameters                       │
│ ────────────────────┐                     │
│                     └──────────────       │
│ ──────────────┐  ┌──────────────          │
│               └──┘                         │
│ Delay  3            Width  15             │
│                          Cancel    OK     │
└─────────────────────────────────────────┘
```

The controls in this box are summarized in the following table.

Delay	This text box allows you to edit the delay time setting of the selected device. Allowable settings are in the range 0 to 32,767.
Width	This text box allows you to edit the width time setting of the selected clock device. Allowable settings are in the range 1 to 32,767.

◆ See Chapter 9, Primitive Devices, for more information on how you can set the initial value of a One Shot device by setting the inversion on the output pin.

Trigger Setup...

This command displays the Trigger Setup dialog, as follows:

```
┌─────────────────────────────────────────────┐
│ Simulation Trigger Setup                      │
│ ┌─────────────────────┐ ┌──────────────────┐ │
│ │ Signal Value Condition│ │ Time Condition   │ │
│ │ Signals ┌─────────┐  │ │ Time  0          │ │
│ │         └─────────┘  │ │                  │ │
│ │ Value   ┌─────────┐  │ │ ○ <  ○ =  ○ >    │ │
│ │         └─────────┘  │ │ ○ Every          │ │
│ └─────────────────────┘ │ ● Always         │ │
│ ┌─────────────────────┐ └──────────────────┘ │
│ │ Actions             │                       │
│ │ □ Beep              │                       │
│ │ □ Stop              │   Cancel    OK        │
│ │ □ Reference Line    │                       │
│ └─────────────────────┘                       │
└─────────────────────────────────────────────┘
```

NOTE: The Trigger Setup dialog can also be displayed by clicking on the Trigger button on the Simulator Palette.

Trigger Conditions

The trigger is activated when two sets of conditions are met:

▩ The time condition—i.e., the current simulator time value—is less than, equal to, greater than, or a multiple of a given value.

▩ A signal value condition is met: i.e., one or more signals are at specified levels.

Signal Condition Controls

The controls related to the signal condition are summarized in the following table.

Signals	In this text box, you can type the names of one or more signals whose values will be compared to the hexadecimal integer value typed in the Value box. One or more signals can be entered, using the following formats: CLK — The single signal CLK D7..0 — The signals D7 (most significant bit), D6, D5...D0 IN1 OUT3 — The signals IN1 and OUT3
Value	In this box, you enter the signal comparison value as a hexadecimal integer. This value is converted to binary and compared bit-for-bit with the signals named in the Signals box. The rightmost signal name is compared with the least significant bit of the value, etc.

Time Condition Controls

The controls related to the time condition are summarized in the following table.

Time	In this box, you enter the time value as a decimal integer. The meaning of this value is determined by the switches below it.

<, =, >	These buttons indicate that the trigger will be activated when the simulation time is less than, equal to, or greater than the given value, respectively.
Every	This time option specifies that the trigger will be activated every time the simulator time equals a multiple of the specified value.
Always	This specifies that the time condition should be considered to be always true. The time value is ignored.

Trigger Actions

When the trigger is activated, any combination of the displayed actions can be invoked.

Beep	Generates a single system beep.
Stop	Stops the simulator immediately.
Reference Line	Draws a reference line at this time on the Timing waveform display.

Stick Signals...

This command allows you to set the "stuck" status of the selected signals. It displays the following box:

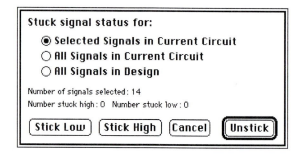

The controls in the box are summarized in the following table.

Selected Signals in Current Circuit	If this option is selected, the Stick High, Stick Low, or Unstick option will apply only to signals currently selected in the Schematic diagram.
All Signals in Current Circuit	If this option is selected, the Stick High, Stick Low, or Unstick option will apply to all signals in the circuit represented in the topmost Schematic window. Only this circuit level is affected, i.e., other open subcircuits will not be changed.
All Signals in Design	If this option is selected, the Stick High, Stick Low, or Unstick option will apply to all signals in all parts of the current design.
Number of signals selected	This displays the number of signals that will be affected by any changes made in this box.
Number stuck high	This displays the number of signals in the selected scope that are currently stuck at a high level.
Number stuck low	This displays the number of signals in the selected scope that are currently stuck low.
Stick Low	This closes the box and applies a "stuck low" value to all selected signals.
Stick High	This closes the box and applies a "stuck high" value to all selected signals.
Unstick	This unsticks all selected signals, allowing them to return to their driven value.

◆ See more information on "stuck" signal values in Chapter 7, Simulation.

Clear Simulation

Selecting this item clears the Timing window (if open); removes all scheduled signal events; sets all devices, signals, and pins to their specified initial values (if any); and recalculates output values for all circuit elements.

Clear Unknowns

Selecting this item clears all flip-flop, counter, and register primitives to the zero state, and attempts to remove all unknown signal values from the circuit. Note that certain circuit conditions may prevent signals from being placed in a known state:

▓ Unconnected inputs that have not be set to a known level.

▓ Storage devices, such as RAMs, that have an unknown stored value.

▓ Any simulation model that does not produce a known output when all inputs are known.

Simulate Menu: Timing Window Commands

Add to Timing

This command adds all selected signals in the current circuit to the Timing display. If any selected items are unnamed or are already displayed, they will be ignored. New items are added at the bottom of the Timing display and will be selected after the add.

Add as Group

This command is similar to the Add to Timing command, except that the selected items are all added as a single group. Where possible, items will be sorted in alphanumeric order with the lowest-numbered item in the least significant bit position.

Auto Add

When this item is checked, any signals added or edited on the schematic will automatically be added to the Timing window.

Remove from Timing

This command removes a trace from the Timing window. If the Timing window is topmost, it will remove all the traces whose names are selected in that window. If the Schematic window is topmost, it will remove from the Timing window any signals that are selected in the Schematic diagram.

Group

This command combines all the selected traces into a single display group. If any of the selected traces were already grouped, they are first Ungrouped and then recombined with other selected items into a single new group.

Ungroup

The Ungroup command breaks all signals in selected groups into individual traces. This command cannot be undone.

Find in Timing

This command allows you to locate a trace in the Timing window by name.

Timing Options

This command displays the following options box:

```
┌─────────────────────────────────────────┐
│ ┌───────────────────────────────────┐   │
│ │ Timing Data Retention             │   │
│ │ ● Retain displayed range only     │   │
│ │ ○ Retain for [        ] time units│   │
│ └───────────────────────────────────┘   │
│                    [ Cancel ]  (( OK ))  │
└─────────────────────────────────────────┘
```

Timing Data Retention

These options allow you to determine how much signal-event data is retained in memory when a simulation is run.

Each time a signal-level change occurs, LogicWorks creates a record in memory containing a reference to the signal, time, new value, and source of the change. In a large simulation, these records can consume enormous amounts of memory. This data can be retained for the following purposes:

■ For use in refreshing the Timing window, should it become hidden and then be redisplayed.

■ For use in Timing window editing operations, such as taking the output from one circuit and using it as stimulus for another.

Note that data can be retained only for signals displayed in the Timing window. Signal-event data for all other signals is discarded immediately after it is no longer required for simulation.

The option "Retain displayed range only" is the normal default, and results in data being discarded immediately after the corresponding point on the Timing display scrolls off the left side. This results in minimal memory usage. The setting is equivalent to entering 0 in the Retain time box.

The option "Retain for x time units" allows you to keep the signal-event data for the specified amount of time after it scrolls off the left side of the screen. If this results in a memory shortage, then the simulation will stop and a message will be displayed.

Show/Hide Timing Palette

This command displays or removes the Simulator Palette. It can be invoked without the Timing window displayed to make use of the current time display and simulation controls.

◆ See Chapter 8, The Timing and Simulator Tools, for more information on the Simulator Palette.

Timing Window Pop-Up Menu

```
Get Info...
Go To Schematic
Remove
Group
Ungroup
Collect
To Top
To Bottom
```

Holding down the ⌘ key while clicking in the label area of the Timing window will display the Timing pop-up menu.

The options on this menu control the display of selected signals in the Timing window, and are summarized in the following sections.

Get Info

For groups, this command displays a box allowing reordering of the signals in the group. This affects the way the combined integer value is shown in the Timing display.

For individual signals, a signal info box is displayed.

Go To Schematic

This command displays the signal corresponding to this trace in the Schematic. If the trace is a group, it shows the corresponding bus, if any, or the first signal in the group.

Remove

This command removes the selected traces from the Timing window. *This cannot be undone!*

Group

Combines all the selected traces into a single display group. If any of the selected traces were already grouped, they are first Ungrouped and then recombined with other selected items into a single new group. This is the same as the Group command in the Simulate menu.

Ungroup

Breaks all signals in selected groups into individual traces. This is the same as the Ungroup command in the Simulate menu.

Collect

Brings all selected items together in the display underneath the topmost selected item. This is used to bring associated signals closer together for easier comparison of timing, etc.

To Top

Sends all selected traces to the top of the Timing window.

To Bottom

Sends all selected traces to the bottom of the Timing window.

DevEdit Menu Commands

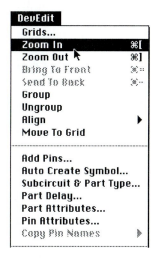

This menu contains all of the DevEditor-specific operations, and is active only when a Device Editor window is topmost.

Grids...

This command allows the user to specify the visible grid spacing and the snap-to grids for objects drawn using the drawing tools.

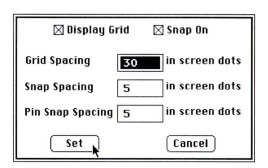

Zoom In / Zoom Out

These commands allow the user to adjust how close-up an object is viewed. The default setting for the DevEditor is to display objects at the same size as they will appear in the schematic. There are five display levels: 1/2 as large, normal, and 2, 4, and 8 times as large.

Bring To Front / Send To Back

These commands are used to set the front-to-back ordering of the selected objects relative to the other graphic objects.

Group / Ungroup

If you select multiple graphic objects (except pins) and then choose the Group command, LogicWorks will treat the group as a single object. To disaggregate a grouped object, select it and then choose Ungroup.

Align

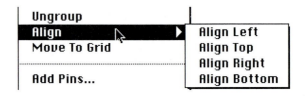

The Align submenu allows you to pick how the selected objects will be aligned. For example, Align Left causes all of the selected objects to be moved such that their left edges are aligned with the leftmost selected object's left edge.

Move to Grid

This command moves the selected object so that its upper left corner will be aligned with the nearest grid line.

Add Pins...

This command brings up the Add Pins floating palette, allowing you to add multiple pins to the DevEditor's pin list.

◆ See Chapter 11, Device Symbol Editing, for more information.

Auto Create Symbol

This command brings up the automatic symbol creation dialog, allowing you to automatically generate a rectangular device symbol.

◆ See Chapter 11, Device Symbol Editing, for more information.

Subcircuit / Part Type...

Displays a dialog that allows you to specify the type of LogicWorks part being created. The LogicWorks types are: No Subcircuit, Subcircuit, Symbol Only, and Primitive.

◆ See Chapter 11, Device Symbol Editing, for more information.

Part and Pin Attributes...

The Set Part Attributes and Set Pin Attributes commands display the standard Attributes dialog for the selected part or pin, respectively. This allows you to set the default attribute values that will be used when the part is used in a schematic.

◆ See Chapter 11, Device Symbol Editing, for more information.

Font / Size / Style

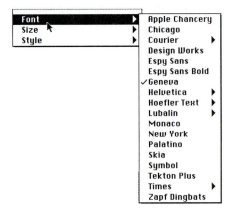

These menu items set the text characteristics for the selected text objects. If no objects are selected, then the text property you set becomes the new default.

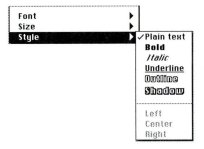

Appendix A— Primitive Device Pin Summary

This appendix lists all the primitive types used in LogicWorks and the allowable numbers and types of pins on each.

◆ For more information on using the Device Symbol Editor to assign a primitive type, see "Assigning a Primitive Type" on page 264.

Schematic Symbol Primitive Types

Primitive Type	Pin Requirements	Comments
SUBCIRCUIT	No restrictions	Symbol having an optional internal circuit. This is the default for symbols created using the DevEditor tool.
SYMBOL	No restrictions	Symbol with no internal circuit.

Pseudo-Device Primitive Types

IMPORTANT: The pin requirements listed in the following table must be followed when creating pseudo-device symbols. These rules *are not checked* by the DevEditor.

Primitive Type	Pin Requirements	Comments
BREAKOUT	Pin 1 is Bus Pin, followed by N Normal Pins, set to Input	Splits signals out of or into a bus.
SIGNAL CONNECTOR	Exactly 1 normal pin, normally set to Input	Used for power and ground connections.
PORT CONNECTOR	Signals—exactly 1 pin Busses—exactly 1 bus pin with any number of internal pins	Makes a connection between the signal to which it is connected and a like-named pin on the parent device.

Simulation Primitive Types

◆ For more information on the simulation primitive types, see "Primitive Devices" on page 189.

NOTE: The following table lists the pin functions and orders for all simulation primitive device types. In some cases, a number of pins can be optionally omitted, so the table gives rules rather than enumerating all possible combinations.

IMPORTANT: In order for a primitive device to simulate correctly:

1) The device pin order *must* follow that given in this table. So when creating the symbol using the DevEditor tool, the pins displayed in the pin list must be in the order described here.
2) The pin type (input/output/bidirectional) must be set appropriately for each pin.

Primitive Type	Limitations	Pin Names and Types	Possible Pin Orders
NOT	Exactly 1 input and 1 output	IN—in OUT—out	1) IN OUT
AND, NAND, OR, NOR, XOR, XNOR	N inputs: 1	$IN_0..IN_{N-1}$—in OUT—out	1) $IN_0..IN_{N-1}$ OUT
X-Gate	Exactly 2 ports and 1 enable	X1 X2—bidir EN—in	1) X1 EN X2
Buffer	N data inputs N data outputs 1	$IN_0..IN_{N-1}$—in $OUT_0..OUT_{N-1}$ —out EN—in	1) $IN_0..IN_{N-1}$ $OUT_0..OUT_{N-1}$ EN 2) $IN_0..IN_{N-1}$ $OUT_0..OUT_{N-1}$
Resistor	Exactly 2 pins	X1 X2—bidir	1) X1 X2

Multiplexer	L select inputs	$S_0..S_{L-1}$—in	1) $IN_{0,0}..IN_{0,M-1}$
	M output bits 1	$IN_{0,0}..IN_{N-1,M-1}$ —in *	$IN_{1,0}..IN_{1,M-1}$ ·· $IN_{N-1,0}..IN_{N-1,M-1}$
	N inputs/ output $2^{L-1} < N^L$ (i.e., the number of inputs per output bit can be less than the number of select input combinations.)	EN—in $^{¤}$ $OUT_0..OUT_{M-1}$— out * $IN_{n,m}$ is the input routed to output m when select value is n.	$S_0..S_{L-1}$ $OUT_0..OUT_{M-1}$ 2) $IN_{0,0}..IN_{0,M-1}$ $IN_{1,0}..IN_{1,M-1}$ ·· $IN_{0,0}..IN_{N-1,M-1}$ $S_0..S_{L-1}$ EN $OUT_0..OUT_{M-1}$ $^{¤}$
		$^{¤}$ An enable input can exist only if $N == 2^L$, otherwise the extra input is assumed to be a data input.	$^{¤}$ Option 2 only if $N == 2^L$
Decoder	L select inputs	$S_0..S_{L-1}$—in EN—in $OUT_0..OUT_{M-1}$— out	1) $OUT_0..OUT_{M-1}$ $S_0..S_{L-1}$
	M output bits 1 AND $2^{L-1} < M^L$ (i.e., the number of output bits can be less than the number of select input combinations.)		2) $OUT_0..OUT_{M-1}$ $S_0..S_{L-1}$ EN
Adder, Subtractor	N output bits N "A" operand inputs required N "B" operand inputs optional 1	$A_0..A_{N-1}$—in $B_0..B_{N-1}$—in CIN—in $SUM_0..SUM_{N-1}$— out COUT—out	1) $A_0..A_{N-1}$ $B_0..B_{N-1}$ $SUM_0..SUM_{N-1}$ CIN COUT * * $B_0..B_{N-1}$ CIN & COUT can be omitted in any combination

D Flip-Flop, D Latch	Must have at least D, EN, and CLK inputs and Q output	S—set in D—D in C—clock in R—reset in Q—out NQ—inverted out	1) S D C E R Q NQ 2) S D C E R Q 3) D C E R Q NQ 4) D C E R Q 5) D C E Q NQ 6) D C E Q
D Flip-Flop with Enable	Must have at least D and CLK inputs and Q output	S—set in D—D in C—clock in E—enable in R—reset in Q—out NQ—inverted out	1) S D C R Q NQ 2) S D C R Q 3) D C R Q NQ 4) D C R Q 5) D C Q NQ 6) D C Q
JK Flip-Flop	Must have at least CLK input and Q output	S—set in J—J in K—K in C—clock in R—reset in Q—out NQ—inverted out	1) S J C K R Q NQ $^{¤}$ 2) S T* C R Q NQ $^{¤}$ 3) T* C R Q NQ $^{¤}$ 4) C R Q NQ $^{¤}$ 5) C Q NQ $^{¤}$ $^{¤}$ NQ can always be omitted * T = J & K tied together
Register	N output bits N input bits 1	$IN_0..IN_{N-1}$—in CLK—in CLR—in $OUT_0..OUT_{N-1}$ —out	1) $IN_0..IN_{N-1}$ $OUT_0..OUT_{N-1}$ CLK CLR 2) $IN_0..IN_{N-1}$ $OUT_0..OUT_{N-1}$ CLK

Counter	N output bits N input bits (optional) 1	$IN_0..IN_{N-1}$—in CLK—in LD—in CLR—in UP—in EN—in $OUT_0..OUT_{N-1}$ —out COUT—out	1) $IN_0..IN_{N-1}$ $OUT_0..OUT_{N-1}$ CLK LD CLR UP EN COUT 2) $IN_0..IN_{N-1}$ $OUT_0..OUT_{N-1}$ CLK LD CLR UP COUT 3) $IN_0..IN_{N-1}$ $OUT_0..OUT_{N-1}$ CLK LD CLR COUT 4) $IN_0..IN_{N-1}$ $OUT_0..OUT_{N-1}$ CLK LD COUT 5) $IN_0..IN_{N-1}$ $OUT_0..OUT_{N-1}$ CLK COUT
Shift Register	N output bits N input bits 1	$IN_0..IN_{N-1}$—in CLK—in LD—in CIN—in $OUT_0..OUT_{N-1}$ —out	1) $IN_0..IN_{N-1}$ $OUT_0..OUT_{N-1}$ CLK LD CIN 2) $OUT_0..OUT_{N-1}$ CLK CIN
One Shot		CLK—in CLR—in Q—out NQ—out	1) CLK CLR Q NQ 2) CLK CLR Q
Clock Osc	Exactly one output pin	CLK—bidir	1) CLK
Binary Switch	Exactly one pin	SW—bidir	1) SW
SPST Switch	Exactly 2 pins	X1 X2—bidir	1) X1 X2
SPDT Switch	Exactly 3 pins	X1 X2 COM —bidir	1) X1 X2 COM
Probe	Exactly 1 pin	PR	1) PR
Hex Keyboard	4 or 5 pins	$X_0..X_3$—bidir STROBE—out	1) $X_0..X_3$ STROBE 2) $X_0..X_3$

Hex Display	Exactly 4 pins	$X_0..X_3$—in	1) $X_0..X_3$
Unknown Detector	Exactly 2 pins	D—in Q—out	1) D Q

Appendix B—
Device Pin Types

Every device pin has a characteristic known as its *pin type*. The pin type is set when the part entry in the library is created and cannot be changed for individual device pins on the schematic.

◆ Refer to Chapter 11, Device Symbol Editing, for information on how to set the pin type while creating a device symbol.

What Pin Types Are Used For

For many general schematic editing purposes, the pin type will be unimportant and can be ignored. However, pin type settings are important in the following cases:

▓ The pin type of each pin is used by the simulator to select what type of output values are generated by a pin. For example, an open collector output will not generate a HIGH drive level.

▓ Pin type information is required in many netlist file formats for FPGA layout and digital simulation.

▓ Other analysis tools may, in the future, use this information for timing and loading analysis.

Pin Types Table

The following table lists the function of each of the pin types available in LogicWorks. The Output Value Mapping column specifies how output values specified by the model are mapped to actual pin drive values.

Pin Type	Description	Initial Value	Output Value Mapping
IN	Input—this is the default for pins created using the DevEditor tool. This setting is used for all pins on discretes except those with some digital function. No output value can be placed on an input pin.	HIGHZ	No output drive allowed.
OUT	Output—always enabled.	DONT01	None.
3STATE	Output—can be disabled (i.e. high-Z). Note: The three-state capability only exists for specific primitive types that have a three-state enable pin. For other types, this will behave like OUT.	DONT01	None.
BIDIR	Bidirectional.	DONT01	None.
OC	Open collector output—pulls down but not up.	DONT0Z	HIGH maps to HIGHZ.
BUS	Bus pin—This does not represent a physical signal but is a graphical representation of a group of internal pins, each having its own type. Bus pins cannot have values and are not supported on primitive device types.	None	None.
LOW	Output—always driving low.	LOW	All values converted to LOW.
HIGH	Output—always driving high.	HIGH	All values converted to HIGH.

Pin Type	Description	Initial Value	Output Value Mapping
LTCHIN	Input to a transparent latch—this is used for calculating cumulative setup and hold times.	HIGHZ	No output drive allowed.
LTCHOUT	Output from a transparent latch—this is used for calculating cumulative setup and hold times.	DONT01	Same as OUT.
CLKIN	Input to an edge-triggered latch—this is used for calculating cumulative setup and hold times.	HIGHZ	No output drive allowed.
CLKOUT	Output from an edge-triggered latch—this is used for calculating cumulative setup and hold times.	DONT01	Same as OUT.
Clock	Clock input—this is used for calculating cumulative setup and hold times.	HIGHZ	No output drive allowed.
OE	Open emitter output—i.e., can pull up but not down.	DONT1Z	LOW maps to HIGHZ.
NC	A no-connect pin.	HIGHZ	No output drive allowed.

Device Pin Type and Simulator Efficiency

Incorrect device pin type settings can have a major impact on simulation speed, even in cases where they do not affect the correctness of the results.

Bidirectional Pins

Using bidirectional pins should be avoided unless specifically required by circuit logic. On primitive types, any value change on a signal attached to a bidirectional pin will cause the device model to be called to reevaluate the device. On subcircuit devices, the simulator must make several passes through all circuit levels that may affect the value of the signal or may be affected by it. Setting a pin on a subcircuit device to be an input or output greatly reduces this overhead.

Output Pins

If a device pin will be used exclusively to drive the attached signal, and the device cannot be affected by changes in value on the pin, then it should be an output type. Changes in the value of a signal attached to an output pin *do not* cause the device model to be called for reevaluation. This is particularly significant for subcircuit devices.

Input Pins

Pins with an input type setting can never place a drive value on the attached signal. On subcircuit devices, this provides an important hint to the simulator that internal value changes on the attached signal will not affect any other circuit level.

Appendix C—
Initialization File Format
(for Windows)

In Windows, you can specify startup options for LogicWorks by creating or modifying an external text file called lw.ini. Use of this file is completely optional: if LogicWorks does not detect its presence, the program will start up with factory defaults. If you choose to create or modify an initialization file, observe the following conventions:

■ The file should be called lw.ini, and should be placed in the LogicWorks directory.

■ Each section starts with a section heading contained within square braces—e.g.: [Drawing].

■ Within a section, each non-blank line is either a statement or a comment. A statement is a keyword (which specifies an option), followed by an equal sign (which is a separator), followed by the option's value. Each statement is terminated by a hard return.

■ A comment is any line that starts with "//".

[System] Section

Modules Directory

ToolFolder = c:\dw\medatools

This statement defines where to look for the external code modules. If this value is not specified, then modules are loaded from the Tools subdirectory within the LogicWorks directory.

Default System Font

> Font = "font_name" font_size [BOLD ITALIC]

This statement specifies the default font which the LogicWorks system will use when no other font has been specified. Certain Tool modules may, by default, display text using this font. If no font is specified, an attempt is made to use a Courier typeface. If no font size is specified then 10 point is used.

"Font_name" is the name of a TrueType font; only TrueType fonts are supported. Font_size is the point size to use. There are two optional style keywords which may be applied, BOLD and ITALIC. For example:

> Font = "Courier New" 10 Bold

Printer Scale Lines

> PrinterScaleLines = None, All, or OverOnePixel

This allows the user to specify whether lines are to be scaled when printed.

"None" is the default. It indicates that no scaling will occur. With this setting, a line's width is printed with the same number of pixels as it is displayed on the screen. When printing to a high-resolution printer (i.e., >= 300 dots per inch), this will cause thick screen lines (busses) to be reproduced as printed lines which do not appear to be much thicker than thin screen lines (signals). This setting is most useful when printing to dot-matrix printers where the printer's resolution is similar to the screen's resolution.

"All" specifies that every line will be scaled so that its printed width appears the same as on the screen.

"OverOnePixel" specifies that lines that have a screen width greater than 1 pixel will be scaled when printed. The result is that signal lines will be drawn very finely, but busses will appear as thick lines.

[System Font Translations] Section

Old_Font_Name = Replacement_Font_Name

Font translations are used when the fonts embedded in a file are not available on the current platform. This section allows the user to define which fonts (available on the current platform) are to be used instead of the specified fonts. The replacement font must be a TrueType font.

Each line in this section specifies a font mapping. For example:

Bookman = Courier New
Times = Times New Roman

…specifies that whenever the font Bookman is displayed or requested, Courier New should be used as its replacement; and whenever Times is displayed or requested, Times New Roman should be substituted.

[Drawing] Section

Initial Directory Settings

Directory = dir_name

This statement specifies the initial working directory. If it is omitted from the .ini file, the working directory will default to the value set by the Windows Program Manager.

Font Settings

XXX_Font = font_name font_size [BOLD ITALIC]

This statement specifies the font for text items appearing in a Schematic document. Font_name is the name of a TrueType font; only TrueType fonts are supported. Font_size is the point size to use. There are two optional style keywords which may be applied, BOLD and ITALIC. The possible items which may have their font specified are:

Default_Font

Attribute_Font, Border_Font, MiscText_Font, Pin_Font, Symbol_Font

Color Settings

XXX_Col =RED, GREEN, BLUE, CYAN, MAGENTA, YELLOW
BLACK, DKGRAY, GRAY, LIGHT GRAY, WHITE

This statement specifies the color for an item(s) appearing in a schematic document. All items in a schematic, except the page background, default to black. The page background defaults to white. The possible keywords for XXX_Col are:

Default_Col

DeviceAttrs_Col, SignalAttrs_Col, BusAttr_Col, PinNumber_Col,
PinNumber_Selected_Col

Device_Col, Signal_Col, Signal_Selected_Col, Pin_Col, Pin_Selected_Col, Bus_Col,
Bus_Selected_Col, BusPin_Col, BusPin_Selected_Col

Page_Col, Boundary_Col, GridMajor_Col, GridMinor_Col, RandomText_Col,
RandomTextFrame_Col

Default Design

DESIGN = circuitName

This statement allows you to specify a circuit file to open when the program starts. This can be used to open a file that is being repeatedly edited, or to open a default "template" file with a standard title block or border.

NOTE: If the circuit is in any directory other than the working directory, a pathname must be specified.

Disabling Untitled Design at Startup

This statement controls the creation of a new untitled design when the program is first started. If the NOUNTITLED statement has a value of On, the program will start up and just display its menus, with no default window opened. The default is Off. This entry will appear as follows:

NOUNTITLED = on

Solid Grid Lines

The SOLIDGRID keyword determines if grid lines are drawn with solid lines. If it is set to "Off," then the grid is drawn with the default dotted lines. On some platforms dotted lines are not correctly supported or may draw slowly. The default entry is:

SOLIDGRID = on

Zoom Factors

SCALES = n1..n11
NormalScale = index

The SCALES statement is used to specify the magnification levels used by the Reduce and Enlarge commands. The keyword is followed by 11 decimal integers, separated by blanks and sorted in ascending order. The 1:1 scale level (at which externally created pictures appear in their original size) is 14. Enlargements are specified by smaller numbers (e.g., 7 gives 200%) and reductions by larger numbers. The default values are:

SCALES = 4 7 10 14 18 24 28 42 63 98 140

NormalScale = 3

The NormalScale statement is used to specify which of the scale steps specified in the SCALES line will be used as the "Normal Size" setting. The index must have a value in the range 1 to 11.

Pin Spacing

PINSPACE = n;

The PINSPACE keyword is used to specify the spacing between adjacent pins when breakout symbols are created. This can also serve as the default for symbols created by other tools. The value must be a single decimal integer, which the program will use as a multiple of the standard grid space of 5 pixels. The default value is 2, which yields a spacing of 10 pixels.

Breakout Parameters

BREAKOUT = dth dtv;

The BREAKOUT keyword lets you control the creation of bus breakout symbols generated by the program. This does not affect any breakouts in existing files, as these symbols are already created and stored with the file.

The BREAKOUT keyword is followed by two numbers for the following parameters:

dth   the horizontal offset (in pixels at 100% scaling) for placement of text names on a breakout.

dtv   the vertical offset (in pixels at 100% scaling) for placement of text names on a breakout.

Disabling "Loose End" Markers on Signal Lines

The NOLOOSEENDS keyword, when set to "on", disables the cross markers that are normally displayed on the screen at the ends of unconnected line segments. The format of the command is:

NOLOOSEENDS = on

To restore the cross markers, use the setting:

NOLOOSEENDS = off

Undo Levels

The UNDO keyword indicates the number of levels of Undo which should be maintained. A value of zero means that there is no Undo. The format of this command is:

UNDO = n

Fine-Tuning Pin Number Text Display

The PINTEXT keyword allows you to adjust the display position of pin numbers on devices. The format of this keyword is as follows:

PINTEXT = dth dtv

…where *dth* defines a horizontal offset for the pin-number text, and *dtv* defines a vertical offset. Both offsets are measured in pixels at Normal Size screen magnification. (See the section above, Breakout Parameters: the BREAKOUT keyword takes the same parameters.) All devices in the design will be equally affected.

IMPORTANT: This adjustment should not be required in normal use and should always be used with caution. No checking is done on the range of these settings.

Changing these numbers in the .ini file *will not* automatically recalculate the positions of pin numbers in existing designs. You can force a recalculate by using the Design Preferences command to change the pin text font or size, then change it back to the original setting.

[Libraries] Section

Library Folder

FOLDER = directory_path

This specifies the folder/directory that will contain the libraries specified in following LIBRARY statements. This statement can be omitted if the libraries are located in the same directory as the LogicWorks executable, or if you prefer to specify a complete library path in each Library statement.

Single Library

LIBRARY = library_path

This specifies a single library to open. The library_path can be simply the name of the library if the library is in the current directory, or a relative path to the library, or a fully specified path from the root. For example:

LIBRARY = lib1.clf
LIBRARY = lib\74LS00.clf
LIBRARY = \mylibs\blocks\controls.clf

All Libraries in a Folder

LIBRARYFOLDER = directory_path

This names a folder/directory to be searched for libraries. All libraries in this folder will be opened. Folders nested inside this folder *are not checked*. The format of the folder name is the same as that for the FOLDER keyword above.

[Timing] Section

The following settings allow the user to control the appearance of all the text and timing waveforms in the Timing window:

Parent_Col = WHITE

Scale_Col = BLACK

LabelText_Col = BLACK

LabelBackground_Col = WHITE

WaveText_Col = BLUE

WaveBackground_Col = WHITE

VerticalLine_Col = GREEN

ReferenceLine_Col = CYAN

HIGH_Col = RED

LOW_Col = BLUE

DONT_Col = LIGHT GRAY

HIGHZ_Col = YELLOW

CONFLICT_Col = MAGENTA

Reference_Font = 12

TimeScale_Font = 12

Wave_Font = 12

Parent_Font = 12

[DevEditor] Section

The options below allow you to customize the look and feel of the DevEditor tool.

Default Font

Font = "font_name" font_size [BOLD ITALIC]

This statement specifies the default font for text items appearing in a DevEditor document. The "font_name" parameter is the name of a True-Type font; only TrueType fonts are supported. The font_size parameter is the point size to use. There are two optional style keywords which may be applied, BOLD and ITALIC.

Grid Settings

GridColor = RED, GREEN, BLUE, CYAN, MAGENTA, YELLOW
BLACK, DKGRAY, GRAY, LIGHT GRAY, WHITE

GridSize = grid

SnapSize = snap

PinSnapSize = pinsnap

The GridColor statement specifies what color to use when displaying the DevEditor's Grid. GridSize, SnapSize, and PinSnapSize are all expressed as multiples of 5 pixels. GridSize specifies the number of 5-pixel intervals between displayed grid lines. SnapSize sets the grid snap for all graphical objects except pins, and PinSnapSize sets the snap used when positioning pins.

Appendix D— Timing Text Data Format

When you Copy or Cut a selected area in the Timing window, two types of data are placed on the system Clipboard:

- A picture of the selected area, in LogicWorks' internal format. This picture is not available to outside applications.

- A text description of the signal value changes occurring in the selected area. This text *is* available to outside applications, and this Appendix describes the text data format.

General Description of Format

The following rules describe the Timing text data format:

- The data is pure ASCII text, with no special binary codes except for standard tab and hard-return characters.

- The format of the data is based on the common "spreadsheet" text data format, i.e.: Each text item is followed by a tab character, except for the last one on a line, which is followed by a hard return.

- Every line has the same number of text items on it.

- The first line of the text (that is, up to the first hard return) is a header which indicates the meaning of the items on the following lines, by position.

- The lines following the header are signal value lines. Each line represents one time step. A complete data line is written out each time any value on the line changes. No line is written out for time steps in which none of the represented signals changed value.

Header Format

The header consists of a series of commands, each starting with a "$", which describe the meaning of the corresponding data items on the following lines.

The header always contains the command "$T" (denoting a time column), followed by a tab character, followed by "$D" (denoting a delay column). The remaining items depend on the traces that were selected in the Timing window.

NOTE: The Timing tool always places the time and delay items in the order given here, although it will accept data with these items present in any order, or even completely missing. Since time and delay are redundant, either one is sufficient. If both are missing, a default delay value will be used.

Single Signal Items

An individual signal is specified by the characters $I (for input) followed by a space, followed by the name of the signal. If the signal contains any blanks or control characters, it will be enclosed in quotation marks.

Grouped Items

Grouped items are denoted by the characters "$I" followed by a blank, followed by the name of the group, followed immediately (without any spaces) by a list of the signals in the group, contained in square brackets. Any group or signal name which contains blanks or control characters will be enclosed in quotation marks.

Data Line Format

Each line following the header must contain one data item for each item in the header line. Thus, the first two items will always be:

■ The time at which the events on this line take place. The Timing tool places in this column the absolute time at which the events occurred (according to the time scale on the diagram). However, when the data is pasted, the times are considered to be relative to the time of the first data line. This is a decimal integer which may take on any 32-bit unsigned value.

■ The delay from this step to the next step. This is redundant information, since it can be derived from the times in the first column. It is provided for compatibility with TestPanel and for improved flexibility in exporting to outside software systems.

NOTE: 1) If the delay and time columns do not match, the longest time is used.

2) The delay on the *last line* has special significance because it indicates the delay from the last signal change to the end of the selected interval. When pasting, this value is used to determine how much time to insert.

The following items on a line will be signal or group values matching the items in the header.

■ Individual signals not in Don't Know or High Impedance states will be either 0 or 1.

■ Grouped signals which are not *all unknown* or *all high impedance* will be specified by a hexadecimal value. The least significant bit of the value corresponds to the rightmost signal in the group list. The special character "X" may be substituted for a hex digit if any one of the four signals represented by that digit is unknown, or "Z" if all the signals represented by that digit were high impedance.

Timing Text Example

The following is an example of Timing text data and its corresponding Timing window.

$T	$D	$I Q[Q0 Q1 Q2 Q3]	$I SI	$I LD	$I CLK
87410	2	F	1	0	0
87412	10	F	1	0	1
87422	10	F	0	0	0
87432	1	F	0	0	1
87433	9	E	0	0	1
87442	10	E	0	0	0
87452	1	E	1	0	1
87453	9	D	1	0	1
87462	10	D	1	0	0
87472	1	D	1	0	1
87473	9	B	1	0	1
87482	10	B	1	0	0
87492	1	B	1	0	1
87493	9	7	1	0	1
87502	4	7	1	0	0

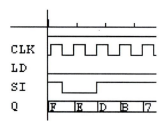

INDEX

▲▲▲ Addison-Wesley
Technical Support

To the Student:

Addison-Wesley provides help for students with installation issues, or if you feel that you have received a defective product. We do not provide assistance with "how to" questions. Please consult with your instructor if you have a question on how to use the software, or if it appears that a particular command or function does not give the expected results.

To the Instructor:

We will be happy to provide assistance to adopters of LogicWorks with any issue that may arise. Please understand that on some occasions we may need to consult with the software developer for answers to specific problems, but we will make every attempt to obtain a fast answer for you.

Before You Call Tech Support:

- Please take time to consult your LogicWorks manual and any release notes that came with the software. These items might provide answers to your questions.

- Are you receiving error messages? If so, when do they occur? What is the exact message? Can you recreate the problem?

- Does your computer hardware setup meet or exceed the minimum system requirements? Please take a moment to compare your hardware against the system requirements on the back cover. Are your hardware and peripherals set up correctly and are all cables attached securely?

- Can your CD drive read the CD-ROM disk appropriately? For Windows users, a quick way to test this is to read the directory of the disk. Use the Windows Explorer or type DIR at the drive prompt (DOS). Do you see files?

Preparing to Call Tech Support:

We will be able to answer your questions more efficiently if you collect the following information before you place a call for help:

■ Specifics of your computer hardware set-up: Processor, internal devices, external devices, RAM, additional cards and OS used.

■ For Windows users: Contents of your Windows start-up files, AUTOEXEC.BAT and CONFIG.SYS. These files are located at the root of your C:\> drive. Use the DOS editor to open and print them out. Many times problems are caused by conflicts between files that are loading at start-up.

■ For Macintosh users: System version, system extensions, and control panels that you have loaded.

■ Know which release of LogicWorks you are using. To find out:

Windows—Choose "About LogicWorks" from the Help Menu.

Macintosh—Choose "About LogicWorks" from the Apple Menu.

Reaching Addison-Wesley Tech Support

Voice: 800 677-6337   Monday - Friday, 9:00 AM TO 5:00 PM, CST
Fax: 847 486-2595
Email: techsprt@awl.com

Software Updates

Updates, patches and extras for this software are available directly from Capilano Computing Systems Ltd. at:
http://www.capilano.com/LogicWorks.